AF540421

HIMALAYAN ECOLOGY

HIMALAYAN ECOLOGY

S.K. Chadha

ASHISH PUBLISHING HOUSE
8/81, Punjabi Bagh, New Delhi-110026

Published by :
S.B. Nangia
for Ashish Publishing House
8/81, Punjabi Bagh,
New Delhi-110026
Tel. 011-23274050
e-mail: aphbooks@gmail.com

978-81-7024-233-8

2022

Rs. 2495/-

Printed by :
New Prints
2026, Rani Bagh
Delhi-110034

Acknowledgement

It is my privilege to thank all those Scholars and Scientists who contributed articles in the present volume.

Words fail to express my sincere regards to Dr. T. N. Khoshoo, distinguished Scientist (CSIR), who encouraged and prompted me to bring out a book on the Himalayan Ecology.

I sincerely acknowledge the assistance extended by Dr. R. N. Sehgal, Associate Professor; Dr. Y.S. Parmar, College of forestry, University of Horticulture and Forestry, Solan (H. P.); Professor R.L. Singh, Ex-Vice Chancellor, Meerut University; Dr. Atul, H. P. Krishi Vishva Vidyalaya, Palampur (H. P.); Prof. S.S. Chib, Head, Post Graduate Department of Geography, University of Jammu, Jammu (J & K State); Dr. M.L. Dewan, Member, National Landuse and Conservation Board, Government of India, New Delhi; Editor, the Hindustan Times, New Delhi; Editor, The Kashmir Times, Jammu; Editor, India To-Day, New Delhi; Shri M. N. Buch, Chairman, National Centre for Human Settlement and Environment, New Delhi and

Shri S. C. Awasthi, an eminent Journalist and many others who very kindly permitted me to reproduce the matter from their publications.

My thanks are also due to the Publishers M/s Ashish Publishing House, New Delhi who extended their full co-operation in bringing out the book in the present form.

S. K. CHADHA

Jammu Tawi

Preface

Himalayas play a vital role in the Social, Cultural and economic life of the people of the Sub-Continent of India. The lofty and mighty mountains influence the ecology and environment. They abound in rare Species of animals, plants and natural resources. The rapid increase in the population of India and the ever increase in the demand of food, fuel and fodder is causing a great threat to the eco-system. The constant deforestation, frequency of Landslides, Snow Storms, avalanches, flesh floods, denudation, over-grazing and exinction of fauna and flora is creating an ecological havoc in the Himalayas.

The present book has been offered to create an environmental awareness among the masses. So that the Sub-Continent may not be deprived of the natural gift of beauty and resources. Of late, the Himalayan States have been undergoing considerable development. There is a dire necessity to incorporate ecological and environmental considerations into the developmental projects to ensure that the eco-system of the Himalayas is not disturbed.

With the above referred Considerations, the present book has been edited to reflect the eco-system of the Himalayas. The various articles in the present volume diverse aspects of Himalayas ecology and environment.

It is desired that the book may prove useful to all those who are interested in the study of ecology and environment of the Himalayas.

S. K. CHADHA

Jammu Tawi (J & K State)

Contents

Contributors

Chadha, Mrs. Kiran, 5A-Lajpat Nagar, Canal Road (Near J & K Bank), Jammu Tawi.

Chadha, S. K., Prof. and Head Deptt. of Geology, Govt. G. M. Sci., College, Jammu Tawi.

Chadha, Miss Veenu, 5A-Lajpat Nagar, Canal Road (Near J & K Bank) Jammu Tawi.

Chakravarti, P. K., Reader, Deptt. of Geography and Applied Geography, University of North Bengal, Darjeeling.

Devendra Pal, Scientist, Wadia Institute of Himalayan Geology, Dehra Dun.

Dewan, M. L., Member, National Landuse and Conservation Board, Govt. of India, New Delhi.

Negi, R. S., Scientist, Wadia Institute of Himalayan Geology, Dehra Dun (U.P.).

Raina, J. L., Prof. of Geography, Govt. G. M. Sci., College, Jammu Tawi.

Rana, R. S., Managing Director and Advisor (Horti.) to the Govt. of Himachal Pradesh, Simla.

Sahni Mrs. Kamal, 6/283, Sant Colony, Bahadurgarh (Haryana State).

Singh, Kartar, Scientist, Institute of Rural Management, Anand.

Talwar, Y. P., Scientist, Coal Survey Laboratory, RRL, Jammu Tawi.

1

Ecological Havoc in Himalayas

DR. S.K. CHADHA

The Himalayas are amongst the youngest mountains in the world. They provide a variety of natural resources to the Indian Sub-Continent, including the life giving water. These resources have been threatened largely by human negligence and activities. In order to regenerate degraded hill environment scientific hazard-management, involving hazard zone mapping and banning engineering development activities in such zones, drainage control and dewatering measures, protection of slopes through vegetation, modification of slopes where necessary, application of engineering technology and improved agronomic practices should be taken up on priority basis.

The problem of landslides in the Himalayas are complex and defy simple solutions. These occur due to the combined effect of intensive human activity, excess rainfall, cloudbursts, flash floods, snow storms, windstorms, fluctuations in the level of subsurface water and slope undercutting by meandering rivers, deforestation and seismicity of the region.

Uttarkhand has been facing the wrath of angry Himalayas since the seventies and more than 1,200 people and thousands of animals have lost their lives. One thousand acre of fertile land has been rendered vulnerable to human. The brutal and inhuman treatment meted out to these mighty mountains is fully responsible for this sorry state of affairs. With barren traits of land widening day by day and depletion of the greenery, the ecological state has altogether changed.

The quantity of the drinking water has gone down drastically and alternate arrangements are woefully lacking. This has further created ecological havoc.

The rainy season in the entire belt of the Himalayas Symbolises Calamity and Chaos to the inhabitants from Assam to Kashmir. Cloud bursting, landslides and slipping down of the huge boulders, masses of debris, broken rocks, tree trunks, etc., has been on the increase. As a result, people living in the hilly regions of Himalayas are being forced to be aliens to each other, due to road blockade or breaking up of the Kachcha roads at several points which ironically are the only arteries, which connect the adjoining villages of the hills.

According to a report by Mr. Mahipal, an eminent Journalist, on July 17, 1986, sixty villages of Chamoli tehsil were ravaged by landslips and landslides. In this natural havoc more than 14 men lost their lives and more than 200 animals perished in a jiffy. Most of the fertile lands were covered by debris and almost all the weak structures and mud houses crumbled.

The torrential rains, strong wind storms and cloud bursts took a toll of 13 people in Tehri Garhwal. It is also believed that more than five people were swept away by the fierce flow of the water. In a similar manner, six people died at Bageshwar tehsil in Almora district. In the Bageshwar tehsil the intensity of landslips was even greater and about half a dozen people died and an equal number were wounded. Such examples can be multiplied from several other areas of Kumaon and Garhwal Himalayas.

Landslides are very common on the Jammu-Srinagar national highway. Every year during the rainy spell landslides block the road near Kud, Batote, Ramban, Banihal, etc. and vistors to Kashmir are Stranded for days. Landslides have also been reported from the eastern Himalayas, Himachal Pradesh and other parts located all along the Himalayas.

In higher attitudes the snow storms and avalanches cause a great loss to human and animal lives. In the Ladakh region of Kashmir and Spiti in Himachal Pradesh dozens of people are buried every season in the avalanches.

The Himalayas are not as strong as they look. Dynamics of change in the entire range is inextricably intertwined with factors such as climate, geology, fauna, flora, water resources, etc. According to Prof. K.S. Valdiya, the Himalayas are made up of weak deformed rocks, which easily give way to the onslaught of rains, shocks of explosions and earthquakes and the vibrations generated by the movement of heavy vehicles and human interference. And due to the precarious balance even a small disturbance triggers changes that rapidly assume alarming proportions.

Unplanned urban development along the length of the Himalayas compounds the problem that concerns not only to 45 million people in the Himalayas and nearly seven times as many in the plains striking the mountain chains but in a sense the future of the entire region. What makes the Himalayas fragile ? Firstly, the geomorphic features due to technically displaced, folded as well as crumpled rocks. Secondly, an intense amount of displacement in these rocks. Thirdly, the unstable character of the Himalayas, the region is subject to periodic earth tremors or earthquakes. Fourthly, some parts of the Himalayas receive exceptionally heavy rainfall which is indeed a vital factor in causing the landslides. Fifthly, the Himalayas have been robbed of its forest cover, specially during the last 25 years, as clearly revealed by aerial photographs and satellite pictures. The deforestation has brought misery and suffering. The fuelwood and fodder are rapidly disappearing and the womenfolk have to travel a long distance and spent more time to collect fuel. And last but not least the instability brought about by large scale construction of dams, tunnels, roads, towers, tanks, ropeways and other projects. The Himalayas are dying with the sound of the dynamite as well as indiscriminate mining and quarring. The ecological blight has taken a toll not only of the quality of lives of individuals but also of the environment. Every tree that is felled now ultimately damages the life support system the soil water and the climate of the Indo-Gangetic plain.

Essentially, the Central Himalayan ranges are like a gigantic sponge. The forests through roots, leaves and mosses catch rain-water and release it gently into the ground for storage in aquifers—natures' own cave like storage tanks—which ration the water in trickles into rivulets supplying the Ganga and the Jamuna. But because of the denudation of mountain slopes most of these aquifers are now dry. During the rainy seasons the water simply cascades down the slopes into river valleys, carrying with it huge amounts of topsoil causing landslides and flash floods. In the dry weather, because the storage system is depleted, there are droughts.

Man is the "agent of change" and many areas of Himalayas have been robbed of the protective vegetable cover. The slopes without vegetation could not be expected to hold soil cover together. Widespread soil erosion is the natural consequence. It has been estimated that nature takes nearly 1000 years to produce a few centimetres of top soil but destabilizing forces of nature in the Himalayas wipe away millions of cubic metres in no time. The rate of erosion in the catchment area of the Himalayan rivers has increased five fold in the geological time scale, the present rate being upwards of 1 mm per year.

Prof. K.S. Valdiya, an eminent Scientist of Nainital (U.P.) is of the view that the rivers are carrying sediments from the Himalayas at the rate of 16.5 hectare-metre per hundred sq. km of the catchment area annually, leading to rapid siltation of reservoirs and lakes. It has also been estimated that the eroded debris carried by the Himalayan rivers have created a new landmass about 50,000 sq. km. in area, extending to about 700 km into the sea.

With a rapid increase in the density of population and urbanisation in the Himalayan areas, the network of roads is increasing. Today the network of Himalayan roads is over 40,000 kms. Some of the new roads have been made at an altitude of 5,000 metres. Khardungla is the highest motor road in the world with an altitude of 5,600 m, surrounded by snow covered mountains. Due to 40°C temperature of the area, it remains open for a few months in a year. The 434 km road from Srinagar (Kashmir) to Leh town has cut down the Journey from 16 days to 2 days but it has made the entire belt unstable and weak.

Proper planning and methodical construction is necessary for ecological balance but when the projects have to be completed in a specified time and no attention is paid to topography, geological structures, drainage patterns and slope studies, the direct result is the ecological havoc. It results is chronic landslips, mudslips and landslides. At many landslide sites, the debris clearance may well be of the order of 5,000 tonne annually.

The construction of dams in this ecologically fragile region is further creating havoc. The construction of the Tehri Dam on Bhagirathi is likely to swallow the entire Tehri town. The dam projects over the Ganga and its tributaries in the hills alone number 22. Apart from deforestation. excavations and resettlement problems it will create instability in the rock formations and there is apprehension of frequency of earthquakes, landslides, soil sinking, soil flowage, soil slip, changes in the underground structures, etc. The continued quarring, mining, installation of transmission lines, towers, microwave, T.V., etc shall heavely damage the slopes and areas environment.

To accommodate the influx of tourists, every year many new hotels and buildings are constructed in Shimla, Kashmir, Nainital, Mussoorie, Darjeeling, etc. This has led to pollution of diverse types, *i.e.*, noise pollution, water pollution, air pollution and garbage pollution. The cases of Dal Lake in Kashmir and Nainital Lake in Uttar Pradesh are very alarming, because more and more sewer pipes have been emptying into these lakes. To promote tourism, ropeways are being introduced in many hill stations like Nainital and Mussoorie. This has made the hill slopes unstable and fragile. Some of the constructions are coming up on old landslides without adequate pretreatment and investments on hill side stability, compounding the problem.

The Himalayan rivers are adding fury to the fire. They are flowing very fast and transport thousands of tonne of sediments every year. The river slopes have become naked, huge landmasses have rolled down into rivers damming them and avalanches of mud and water have uprooted the trees on slopes. Some of the Himalayan rivers are in spate during the monsoon period and cause great loss to life, property and environment. In Sikkim Himalayas, rainfall is often punctuated by flashes of cloud bursts. In hill regions, a cloud burst comes with a speed of thunder, lasts from a fraction of a minute to as long as three hours at a time and leaves behind a tail of devastation worse than that inflicted by the combined effect of rainfall for the rest of the season.

The commonly employed corrective measures against landslides include the construction of retaining walls, benching and grading of slopes, regulating sub-surface water checking erosion of hill slopes, grouting the weaker zones, using binding substances for loose soils, Afforestation schemes should be introduced to conserve soil and water in the hill regions. The vegetation provides benefits of evapotranspiration elimination of which would almost certainly tend to raise the groundwater table. The elimination of root system takes away the suction effect in soil which otherwise offers stabilising effect.

The hills which were once considered the most stable place to live in ironically the same hills have become unstable and unworthy of human living. The malaise lies in our thinking and attitude towards the issue of ecology and environment. Till to-day we have seen the problems of the hills something alien from threat of the plains. An attempt was never made to understand that a close symbiotic relation exists between the hills and the adjacent plains. The hills through the rivers have shaped the plains and plains through their economic activities have modified the eco-system of the Himalayas.

Of late, the Himalayan region has been undergoing considerable development. There is a paramount need to incorporate environmental and ecological considerations into the developmental activities to ensure that short term gains are not allowed to ruin the long term gains. The urgent needs of economic development and the imperatives of environmental conservation need to be meshed with each other by the planner and the policy maker, the development administrator, the specialist adviser and the general public.

If close coordination among the planners, administrators, scientists, policy makers and the natives will not be maintained then there would be confusion in all the programmes of eco-systems. To promote the schemes of eco-system, we have to be on elert and see that every item of ecological and environmental preservation is carried through successfully. The protection and preservation of eco-system is of utmost importance. We have to remember the words of Shri Rajiv Gandhi, Hon'ble Prime Minister of India, who while sending his message on the occasion of foundation stone laying ceremony of the Carton Plant at Pragati Nagar in Himachal Pradesh by Shri Virbhadra Singh, Hon'ble Chief Minister of Himachal Pradesh, on 15th June 1988 expressed, "when the environment is not protected, damage to the environment will extract its price from those living in the vicinity, from others at a distance or even from coming generations."

2

State of Himalayas : A Call for Action

M.L. Dewan

The Himalayan Eco-System is one of the most important and most threatened of life support systems on earth. In the shadow of the Himalayas, live more than 150 million people in India, some of them poorest in the world. The rivers which arise in the Himalayas and flew down into the wide Indo-Gangetic plain support the essential agriculture which sustains these people. The population pressure on land and its mounting demand for fuel and fodder have already denuded areas of the Himalayan Ranges—the life line of India. Himalayan land resource degradation has been a matter of considerable concern in recent years. The Himalayas, our sacred mountains, are in a very poor state of health due to deforestation, increasing pressures of human and animal population and their needs and greeds. The rate at which the forests are disappearing is much faster than the rate of afforestation.

The Government of India and the State Governments are making efforts to conserve the Himalayan environment, but are also realising the fact that they alone cannot undertake to carry out such a gigantic task.

Work by the Ministry of Environment and Forests, including National Wasteland Development Board, Soil and Water Conservation Division of the Ministry of Agriculture, Department of Non-conventional Energy Sources and other related Government Departments are still not able to carry out more than a fraction of what is needed in the Himalayas. It is urgent that people's participation must be promoted for such a Himalayan Eco-system Programme.

In the call for action, six pronged action programme which in brief asks for increased tree plantation decreased deforestation, improving springs and water harvesting, promotion of renewable sources of energy, integrated programme for improving crops, livestock, agro-forestry, grassland, biofencing, etc. and intensification of poverty alleviation programmes, etc. The strategy also includes identification of critical catchments and their repair, as also an urgent call to the people and to the government for a concerted action.

The proposals for accelerated action identify :

(*a*) The problem

(*b*) The success stories

(*c*) Proposals for accelerated action

(*d*) Recommendation for Five-Year Programme (1987-92) including investment needs.

The "State of Himalayas : A call for Action" has been prepared by Dr. M.L. Dewan, Member (Soil Conservation), National Land Use and Conservation Board, on invitation of Ministry of Agriculture—Land Resources Commissioner. The emphasis is on people's participation, first report of which was published in 1985. This followed with several programmes—eco-development camps, nurseries, plantation, formation of local conservation and forest societies in the Himalayas.

Principle Behind the Action Plan

First and foremost, it has to be accepted by all concerned that the best and the most practical route for conserving natural resources (soil, water, vegetation, etc.) is through their wise and rational utilisation. The local communities place a value on the resources; it is precisely because of this that they are utilising them and are benefited by the use. What is important is not to alienate the local communities by dropping a "legal curtain" between the people and the resources but to find ways and means of seeking people's participation for sustainable use of the resources. This plan, therefore, places the PEOPLE at the centre of the stage and calls for management of the resources through them and by them.

The *objectives* of the Call for Action are :

(*a*) Through people's participation achieve :

(*i*) the protection of soil;

(*ii*) the promotion of employment of local communities who should be the partners in conserving and utilising the natural resources in a sustainable manner;

(*iii*) acceleration of income and the improvement in the living standards of the people;

(*iv*) development and improvement of the availability of surface as well as ground water resources;

(*v*) promotion of viable and sustainable farming systems which are integrated in nature and serve not only the comprehensive needs of the farming communities but are also environmentally sound.

(*b*) To ensure that floods are prevented thus contributing to the development of sustainable agriculture in the low lands. The reduction of siltation of the rivers has manifold other benefits.

(*c*) Conservation of the uplands and the catchments will prolong the life of the major and minor reservoirs that are already constructed.

The Frame-work for Action

In view of scarce resources, the focus should at least for the coming five year period be on the most severely degraded areas which are also most densely populated. A survey is being undertaken (using satellite imagery and aerial photography) which aims at identifying and specific villages and the lands around them that need to be developed. Many of these catchments are identified. NGOs and voluntary agencies undertake leadership in each catchments, tasks proposed and 6 pronged action plan outlined. After detailed action plan, the task is to act.

Twenty five *critical* watershed are being selected for urgent and concerted action. Non-government and people's organisations will give greater attention to these critical watersheds which are identified as giving greater damage in flooding, erosion and siltation. Government can continue its plan but in whatever way it can help the local societies it will be extra effort. These 25 critical catchments are based on detailed soil surveys as well as use of remote sensing imagery and aerial photographs. These are the critical catchments which are defined as "where the quantitative studies have shown high values and needing urgent attention due to high erosion evidence by the following data :"

(*a*) Sediment yield index weightage values use (potential soil detachment).

(*b*) Delivery ratio representing percentage of silt detached reaching the reservoir site (or potential reservoir site).

(*c*) High slope percentage, shallow depth of soil, texture of soil, soil calcareousness, gravelliness, soil cover, land surface/characteristics, existing erosion conditions such as gully, ravine and landslide conditions, etc.

These 26 critical catchments are in the following States ;

(*i*) Arunachal Pradesh	1
(*ii*) Assam	1
(*iii*) Himachal Pradesh	6
(*iv*) Jammu and Kashmir	3
(*v*) Manipur	1
(*vi*) Sikkim	2
(*vii*) Tripura	1
(*viii*) Uttar Pradesh	5
(*ix*) West Bengal	1
(*x*) Other N.E. States	2
(*xi*) Others not yet clearly defined	3

Some of these critical catchments need more studies before delineating them exactly on the map and on the ground.

The "State of Himalayas : A call for Action" is based on six pronged strategy as follows :

1. Increase Production and Planting for Trees

(*a*) Regeneration

(*b*) Planting programmes of fuel, fodder, fruit and multipurpose trees.

(*c*) Bio-fencing/social fencing.

2. Reduce Consumption of Woodfuel and Cutting of Forests

(*a*) Alternative Energy (Solar/biogas/wind, etc.)

(*b*) Mini Hydel

(*c*) Smokeless chulhas

(*d*) Other measures for energy conservation.

3. Special Treatment of Critically Damaged Areas

(*a*) Select critical micro/mini catchments

(*b*) Treatment around villages/towns, where damage is extensive

(*c*) Landslides must be treated by special means (in situ planting Juniper, other evergreen, etc.).

4. Revival/Regeneration of Springs and Ponds/Water Harvested in Small Scale Water Dams+Mini Micro Hydel+not Large Dams.

5. A Call to People

(*a*) Family planning is *Most Urgent*. Produce less children.

(*b*) Take the challenge. Be self-reliant.

(*c*) Establish local conservation-societies, for survival, work cooperatively and not always depend on the government, District Forest Association, People's Nurseries, catchment development board.

(*d*) Organise into special sectors—special groups such as :

(*i*) Ex-Servicemen

(*ii*) Students

(*iii*) NGO/voluntary groups

(*iv*) Women

(*v*) Other youth

6. A Call to the Government

(*a*) Help those below poverty lines.

(*b*) Promote integrated farming.

(*c*) Utilise wealth of NGOs, voluntary organisations; military, paramilitary, border road organisations, ex-serviceman.

(*d*) Promote people's and school nurseries.

(*e*) Assist in creating catchment development boards, district forest associations, village van panchayats.

(*f*) Undertake training programmes in all aspects of Himalayan Eco-Development.

(*g*) Speed up the creation of a research, development and training centre in the Himalayas like the one proposed Centre for Himalayan Conservation and Development.

(*h*) Decentralise decision making process.

(*i*) Have a Central Coordination Cell to monitor and evaluate the ongoing Programme in the Himalaya.

While considerable financial, moral and resource support is needed, groups like Himalayan Eco-System Development and Plantation, Utilisation and Reforestation Enterprise (PURE) combining retired UN Forest, Army Officers and some young people and many many other groups of people throughout the Himalayas are determined to save the Himalayas mainly through people's effort. The task is tremendous and at present being done on an ad hoc basis. A Central Coordination Cell whose objectives are mainly to stimulate, catalyse, monitor and evaluate projects is urgently needed. It is also proposed that a systematic and coordinated effort for eco-development in the Himalayas should include a computerized system which will indicate the inputs of the people's groups, the government's as well as other organisations. A Himalayan Conservation Research and Development and Training Centre as planned needs to be expedited. Himalayan Eco-Development Policy Board should review the progress at least once a year. The Himalayas are in critical stage of deterioration and only a concerted and integrated action by all concerned/interested is urgently needed. To save the Himalayas is to save Indo-Gangetic plain and in effect to save the country from an eco-disaster.

The appeal for funds will be for each critical area and this appeal is specially directed to the industry for each project. It is hoped that about Rs. 5 lakhs are granted for each project which should spell out the area, the problem and the action planned with quantitative details. Any interested industry is most welcome to enquire about the details indicating the area or the region or catchment they may be interested in the what maximum contribution they can make in a given time frame (say two years in the first instance). Details can then be worked out by the National Land Use and Conservation Board/or the Member (Soil Conservation) and associates for the Action Programme. It can have a district (sub-district) approach or a catchment (micro/mini) approach, but the area of operation must clearly be defined. The kind support of all is solicited. The target for 2 years, *i.e.*, 1987 and 1988 for fund collection is a humble amount of Rs. 10 lakhs from people, Rs. 30 lakhs from the industry and Rs. 60 lakhs from the government. Once this target is achieved, the speed of action will be increased depending on the resources, financial, human and others. This is, therefore, an appeal to help and to cooperate.

3

Horticulture : Vital Role in Himalayan Eco-system

R.S. RANA*

The most urgent task facing mankind today is to find a comprehensive solution to the problem of hunger and mal-nutrition through methods that do not over-burden stocks of non-renewable resources such as soil and minerals and do not impoverish the environment.

Agriculture is supposed to meet the entire requirement of food world over, however, "Agriculture is for the plains only while Silviculture is for the hills and mountains because when the plough invades the hills and mountains it destroys the land. . . ." Of the world surface only 8 to 10 per cent is used at present for food production, however, with the aid of trees which can be

*Managing Director, HPMC and Advisor (Horticulture), Government of Himachal Pradesh.

grown in the most unlikely locations, atleast three quarters of the earth could supply human needs, not only of food but of clothing, fuel, shelter and other basic products. The rapidly increasing human population and the wide extent of mal-nutrition call for greatly increased food production coupled with rational management of natural resources. This can be achieved only on the basis of scientific knowledge, which undoubtedly in many fields of biology is at present wholly inadequate. This is particularly true as far as trees and tree crops and their uses in ecological cultivation are concerned.

At present, agriculture in most part of the world is virtually exclusively geared to cereal growing or at best live stock rearing. Cereals which constitute the stapple diet of most of the world's races, demand annual cultivations which are enormously expensive in labour or machinery, requires large inputs of water and fertilizers and are extremely vulnerable to the vagaries of weather. Harvest failure due to drought, flood or storms can lead to disastor.

Live stock rearing as mixed farming is good. However, as it is carried on in many of the advanced countries where it depends on few strains of grass and clover is an extremely unproductive system of food production and can also be disastrous when grazing areas are affected by flood or drought.

In the light of the conspicuous failure of conventional agriculture to fulfil the nutritional need of the rapidly growing population, far sighted agronomists in many countries are turning their attention to the numerous advantages of tree crops. First and foremost, trees offer the possibility of far higher food yields per acre. Whereas the live stock rearing in temperate regions produces an average of about two hundred weight of meet per acre and cereal growing an average of about half a tonne per acre, apple trees can yield atleast seven tonnes per acre. There are also instances where it is reported that some leguminous bean bearing trees like honeylocust produce 15 to 20 tonnes of cereals equivalent. In tropical areas and under conditions of multiple cropping—where trees are inter-planted with vines, vegetables or even with cereals—far higher yields can be expected. There are many other examples where plantations of trees of various species like honeylocust, carob, algarobas, wall-nuts, oaks, peacans, olives haizulnuts and persimons will provide good per acre yield. From this it is evident that the crop yielding trees offer the best medium or extending agriculture to hills, to steep places, rocky mountains and to the lands where rain-fall is deficient.

Advocates of factory farming or synthetic food manufacture might claim that by such means still higher productivity can be achieved than by tree cropping but it must be realised that extensive acreage of cereals and protein crop such as soyabeans, linseed and other crops are required to feed battery-hens and battery-cows. Production of synthetic foods also demand large quantities of oil or coal—non-renewable resources which are not becoming only increasingly expensive but also scarce. As a machine for supplying the necessary factors in sustaining the human and animal life, the trees with its deep ever-questing roots, seeking out the riches of sub-soil and its mass of foliage high in the air, utilising atmospheric minerals and solar radiation by scientific process of photosynthesis is far more efficient than any system devised by man.

Another outstanding advantage enjoyed by tree cropping is that they can tolerate conditions in which every other form of food production would be impossible such as steep rocky hills. Both olives and chilgoza pines for example can be planted in clefts of rocks where no soil at all is present. Trees can tolerate both rarified air of great heights and polluted atmosphere of industrial cities. In recent years, apple orchards have been established on heights of over eleven thousand feet above sea level in Kinnaur and Spiti areas of Himachal Pradesh.

Oleander and many other plants of great importance are found growing in the highly polluted road-side atmosphere in cities as well as in country-sides. In fact the 'tool' with the greatest potentials for feeding man and animals, for regenerating the soil, for restoring water systems, for controlling floods and draughts, for creating more benevolent micro-climates and more comfortable living conditions for humanity, is the tree.

These facts clearly suggest that tree crops are the only answer to solve our problems this can be applied to every part of the earth where trees will grow and animals exist. Such a system is capable of operation on the smallest or the largest scale; it is far less demanding on energy, machinery and irrigation than the conventional agriculture and far from damaging the environment. It conserves and improves both soil and water resources and purifies the atmosphere. It is a system called as tree farming or modern silviculture or modern horticulture.

All over the world at various times and in different forms, the tree farming has contributed appreciably to the man's subsistence and often saved whole population.

from starvation. It was, therefore, natural in former days and certainly until the advent of the industrial revolution in western countries and mass scale colonisation of third world countries that the tree groves should have been looked upon by people generally as useful adjuncts to the farming economy and at that time the forested areas were considered to be complementary to the cultivated arable farms and gardens and no artificial dividing lines or barriers like present time existed.

Several factors, however, have combined during the past two centuries to destroy the self-sufficient subsistence economy of earlier times over the major parts of the world in general and Himalayan tracts in particular. Tremendous increase in population created new demands for extra production of grains which could only be met by more intensive farming. Society changed both in organisation and needs. While this was happening, food production became specialised, breaking up into separate practices and disciplines. Forestry which had earlier been an integral and useful part of agricultural scene was virtually relegated to the role of supplying fire wood and timber only which resulted in mass scale destruction of forests consequently up-setting the balance of natural influences and causing extensive soil erosion, frequent lowering of water tables and creation of unravelled opportunities for spread of plagues and apedemics. This was the time when bulk of the century's scientists, farmers and foresters started looking upon the salviculture as a separate technique and having no possible relevance to the growing of food. The culture of fruit had been allotted to the horticultural sphere and orchards were considered as falling into the domain of garden work. There were few contracts at scientific level between foresters and agriculturists and virtually none in the technical or practical fields. Foresters had withdrawn entirely from the food production industries. Such a state of affairs had most deletarious effect upon whole applied science of tree culture and greatly retarded the development and its contribution towards feeding the people.

Horticulture on the other hand also did not remain free from the deletarious effect of monoculture and could not subscribe to the problem of solving food crisis to the extent it was expected. It could not contribute fully even in the improvement of the ecology because it is commonly supposed that neighbouring plants rob each other of sunlight, air, water and soil nutrients and that therefore, economic crops must be grown in sterlized isolation, with all potential competitors ruthlessly eliminated. However, in recent years, people have been started recognising ill-effects of such a system and the necessity for restoring tree cover and conservation with mixed plantations. New systems of farming are being developed which are described as three-dimensional tree farming; three dimensions being –conservation, tree crops and live stock.

Any general pattern of such farming is to have large belts or blocks of economic trees intercepted with narrow strips of grasses or other herbage along which herds of live stock could move and fed from wood lands and producing meat, milk, eggs, wool and other items. The system forms a natural biological cycle into which man fits perfectly; he can eat the food harvested from the trees and the flesh or produce of the live stock or sell them. The manure of the animals is returned to the soil and encourages the healthy and vigorous growth of plants, thus reducing the need of the fertilizers to a minimum.

This system of farming can also be called as three dimensional forestly and offers more than a system of specifying man's basic needs of food, fuel and other essentials. It offers nothing less than a new way of life which could provide rewarding and purposeful occupations for large populations. Such a farming can provide many highly skilled jobs which could give the ambitious technical-minded young men and women of today status and satisfaction atleast equivalent to any available to the industrial worker and perhaps, it would be the only occupation which could make a considerable contribution in the integrated development of Himalayas.

Tree farming has got a special place in the ecological cultivation particularly in hills because this is the most productive and healthy form of land-use and the one most beneficial to the soil and other factors, in the environment including man. Needless to point out that trees exhale oxygen and the tree groves are called *Nature's lungs* and in conservationists circle much concern has been expressed that the wholesale destruction of forests may lead to a reduction of atmospheric oxygen below what is essential for the world's human and animal population. The well known Forester, Richard St. Barbe Baker who founded *The Man of the Tree Society* in 1922 has stated that for minimum safety a country should have about 30 per cent of its surface under tree cover. In many countries, tree cover is well below this figure.

The monoculture even of dense forests trees does not fit well in the ecological cultivation. The natural ecological climax is the mixed forest or mixed tree groves. It is not a mere conglomeration of assorted plants, it is a highly complex system of checks and balances adapted to the climatic and soil conditions of the area.

In nature, the diverse plants and animal spices do not—nor can they—exit in isolation. The survival of each type depends upon the presence of others. An association of plants and animals known as a *biocoenosis* is formed in giving habitats under environmental conditions. It includes plants synthesis organic substances (producers), animal feeding on these plants (consumers), carnivores and parasites living at the expense of consumers and organisms capable of mineralising organic substances that create conditions favourable for plants. The proportions of the number of individuals of the forest species in the *biocoenosis* are also mutually conditioned.

Man has destroyed forests and has turned vast territories to cultivation but in the process he has broken up the *biocoenic* associations formed in the process of evolution. This has concurrently affected self-regulation of the population of species and man has thus practically taken up himself the concern of balancing the relationship between cultivated plants, animals and their environment. Though a whole armoury of technical means is employed for this purpose but unfortunately only too frequently in the blindest manner and that is why we are now faced with problems of imbalanced agriculture and crop protection together with worsening of environmental conditions and serious erosion.

There are important indications to show that tree crops planted for different purposes can make real contribution to integrated agrocoenoses, fulfilling modern requirements and conserving land health, as well as improving environment in which we live. However, it is believed that in creating such plantations of economic importance, man should conform, as far as possible to the ecological conditions of the region and the aim of plantation should be to approximate the region's natural ecological climax so as a agrocoenoses is formed which would be fully adapted to the local soil and climatic conditions. Thereby a new environment would be created with much of the diverse and abundent vitality of the primeval forest, with its natural checks and balances, its mechanism of biological control, but consisting entirely of economic species carefully selected to perform single or multiple functions for the benefit of the man.

The expression 'multiple usages' as applied to new system of farming may be open to various interpretations. Much depends on how far the person who employs the expression is prepared to go in putting it into practice. Strictly speaking in system of tree culture (may it be called as orthodox silviculture; modern horticulture; forest farming; three dimentional forest farming or even permanent agriculture) which extends the scope of forest plantings beyond the normal work of supplying timber could be justifiably defined as one of multiple usages. Within this sphere of activity as various abovementioned names imply, there will, of course, be numerous divergencies. It is heartening to note that there is now every expectation that tree cropping designated by different names will once more attain to significant force in the hill agriculture economy. The horticulture plantations going on in Himalayan hills, may it be Kashmir, Himachal Pradesh or U.P. right extending upto the exterior areas in east, clearly indicate that there is much awakening towards this method. Moreover, modern multiple usages method of tree cropping are vastly superior in capacity for sustaining output than were the old haphazard ways of horticulture. The putting of waste land to additional productivity within the general pattern of multiple usage not only means following approved ecological principles of land health but it also brings in appreciable financial advantages. Experiences of horticultural development work in Himachal Pradesh clearly indicate that with a view to keeping up with the pace of horticulture development and the challenges of tomorrow, we need bringing out changes in orchard composition and practices thereof. In the past much emphasis has been laid on monoculture of single fruit and as a result the growers have started facing big problems. Plantation should now be stocked with different species so as unlike present plantation it becomes a multistorey organism comprising both low-growing and high-growing trees with canopys of different levels, light demanding and shade tolerant species; low shrubs and carpet of small plant and even mycorrhizal fungai, all co-existing harmoneously and each making its individual contribution to the energy and productivity of the hills. Interestingly in such plantations the layering is rooted below ground also; the roots of the highest trees peneterating to the deepest strata of the sub-soil; those of smaller trees and bushes occupying intermediate layers, while shallow-rooting annual and perinnial plants send out a mat of roots just below the surface.

For the improvement of Himalayan economy, it would be necessary to have a full integration between horticulture and silviculture and from the outlines of the development of tree farming mentioned in this study it would be clear that the aim of the system is to increase and diversify the productive capacity of our land so that instead of only one item, may it be timber or food, their output should also include a wide range of raw materials. The proposed Agro-silviculture in its broadest sense defines all plant culture live-stocks keeping as pack of once whole biological cycle looking up each farm unit as a progressing entity. Forestry is integrated with farming, animal husbandry and horticulture to achieve both maximum output and optimum conservation of the given area. Strictly speaking when such a system is fully applied in practice it becomes three dimentional. Such a system if ever can be practised, will not only liberate the area from economic inertia but will also lay down modelities for hill development planning which is as yet a little understood subject in the annals of development. From the lessons learnt from the past work it would be more appropriate if the need of the areas in Himalayas are looked after more intensively in future. It is further suggested that future programmes for this area should be integrated development programmes including all aspects of plantations and building up of infra-structure for supplying of package of inputs and other necessary services simultaneously. It is also suggested that water shed should be the primary planning unit rather than an administrative unit like State, District or Block etc. Needless to point out that the water shed approach would provide a sound base for programming for plantations, soil conservation, water harvesting and harnessing and land use planning etc. It is possible to deliniate water shed into micro and mini water sheds and to prepare mini plans of the forest farming according to the set norms and designs.

4

Horticulture : Vital Role in Himalayan Eco-system Development

M.L. DEWAN

The Himalayan eco-system is one of the most important and most threatened of 'life support' systems on earth. In the shadow of Himalayas live more than 150 million people—some of them, the poorest in the world. The rivers which arise in the Himalays flow down into the plains and support essential agriculture which sustain the people in recent years, these rivers are turning into rivers of sorroow, periodically flooding the plains and taking toll of life and property. The water-sheds in the Himalayan eco-system are being widely devastated—by clearance for agriculture on hills, by logging and cutting for fuel, by overgrazing and by badly managed road building. Poorly managed human activities in the Himalayan terrain are causing accelerated erosion with disastrous consequences. In India, we are losing over 6,000 million tonnes of top soil every year. The eroded soil takes 8.4 million tonnes of soil nutrients along, in addition to silting the reservoirs and river beds and damaging crops, houses and public utilities worth Rs. 1,000 crores every year by floods. Population pressure on land for want of fuel and fodder has already denuded large areas of the Himalayan ranges—the life-line of India. The rate at which forests are being lost is much faster than the rate of afforestation.

Land Resources

Part 1 : Himalayan Mountain Area

Name of the Land Resources Region	*Area in Sq. Kms.*	*Percentage of Land Resource Area (India)*
1. Northern Himalaya Snow-clad Region	1,16,000	3.56
2. Northern Himalaya Alpine Grass Meadow Region	98,250	3.02
3. Northern Himalaya Forest Region	1,31,750	4.08
4. North-Eastern Himalaya Alpine Grass and Meadow Region	16,000	0.50
5. North-Eastern Forest Region	1,61,000	4.97
TOTAL :	5,23,000	16.13

Part 2 : Himalayan Sediment Area

The areas which are directly fed by the Himalayas rivers and which form the Gangetic plains, the sediment coming from Himalayas are as follows :

A. Punjab-Haryana Aluvial Plains Area	1,01,250	3.10
B. Upper Gangetic Alluvial Plains Region	2,00,000	6.15
C. Lower Gangetic Alluvial Plains Region	1,45,500	4.41
D. Assam Valley Region	88,500	2.73
E. Rajasthan Desert Region	1,91,000	5.87
	7.26,250	22.26
Grand Total of Himalayan Region	1,249,250	38.39

The Himalayan Eco-system Programme would be a comprehensive attempt to protect the production base in the Indian Himalayas and plains by reducing soil loss and sedimentation, and by preserving the ecological balance. Its task is to recognise and resolve the problems and the threat to the fragile eco-system through firm action and to protect the watershed with the full involvement of the local people and all others interested. On this depends the life culture and civilization of the Himalayas and the Indo-Gangetic plains. This calls for urgent time scheduled measures.

Three tasks of the Himalayan Eco-system Programme are :—

(*a*) to develop and apply or stimulate application of a technical package of practices for watershed management;

(*b*) to lead to development or strengthening of an appropriate infrastructure;

(*c*) to further improve the social, political and public policy on environment in the Himalayas.

Technical Measures for Conservation and Development

The measures for conservation and development will depend on the land resources whether it is in Part I or Part II of the areas indicated above. In principle, for the Part I, the following technical measures will be required :

(*a*) *Fencing, preferably social fencing* and control for *Regeneration* of vegetation in the Himalayas.

(*b*) *Tree Planting and Afforestation* : Large areas of the Himalayan ranges are already deforested due to population pressures and the needs for fuel, fodder and construction needs of the people. The forest contractors and others are also felling trees. In addition, the shifting cultivation system that is being followed by many jhumias are also depleting the Himalayas of their forest resources. A large scale tree plantation programme is essential not only to replenish what is lost but for increasing fuel, fodder and other economic and food needs of the people such as Horticulture species.

(*c*) Water harvesting and reservoir construction and maintenance.

(*d*) Terraces, check dams and gully plugs, other engineering structures.

(*e*) Agronomic, grassland pasture development.

(*f*) Socio-economic measures including people involvment for conservation and management of the soil.

It is in this context of Himalayan Eco-system development that I believe, horticulture has a vital role to play. The following new ideas in the field of horticulture development in the Himalayas should prove useful :

(*a*) *Production of Olives in the Inter-mountain Valleys* : There is some evidence that some wild olives exist in certain areas of Kulu and Manali in Himachal Pradesh. Some attempts have been made to introduce important varieties or olives in these areas, but, apparently no systematic or large-scale efforts have still been made. The role of olives in the oil production and its value in the food for the growing population, both of mountain and plain areas, is well-known. It is in this context, large-scale olive production in suitable areas of the Himalayas need a great promotion including planting material from Italy and other Mediterranean countries and thereafter planting care is also important to introduce some olive plant nurseries on a large-scale.

(*b*) *Introduction of Japanese or Chinese Mandarin* : Also called as Conquats—in the low hilly areas of the Himalayan ranges where already some citrus grows, is very much worth exploring and the introduction of these plants should be promoted.

(*c*) *Fruit Tree Plantation Brigades* : In addition, although apples have taken well in most of the Himalayan valleys, there are many other temperate fruit trees that need to be further promoted and planted on a large-scale. The planting of these trees is being done in certain valleys and certain areas by some entrepreneurs, but, much more is to be done. I am, therefore, suggesting a few procedures for large-scale involvement of youth as Fruit Trees Plantation Brigades.

(*d*) *Allocation of Certain Areas of Plantation to Young Entrepreneurs* : They would then be self-employed and carry out the work not only of planting and nursing the orchards, but also, later on utilisation as well as processing, preferably on a co-operative basis depending on the areas involved. It is this context that I suggest that each one interested should take up small area and develop his or her own orchards. Develop your nursery and through your nursery, help the small farmers to grow more fruit trees and then take care of these orchards. Each tree being a living entity will respond to your care and will produce lot more than the average of the area.

Marketing : It will still be a major problem and also a pressing one. Here you have to work co-operatively as a group with an interest. Make sure that you get the fair price for your commodity. Market them in a systematic and regular way. Marketing is both a science and an art and this you will have to learn. Horticulture industry is important. One of the things which you will have to develop much more and you will need to develop is fruit concentrates which can then be cheap. Then transport these to the big cities where the market is great and somehow try to replace the western type of drinks.

Fruit Juices for the Masses : Introduce large-scale cheap fruit juices which are prepared at big centres by adding appropriate quantity of water (with or without ice) to these fruit concentrates. At present the fruit juices are available at certain places, but at a high price which most people cannot afford. You can provide it at half the cost or less of normal aerated drink. Many of those who cannot afford it now may be able to afford the cheap and nutritious and tasty genuine fruit juice.

Fruit Revolution : In addition, fruits only in limited quantities or limited types are available for the masses who cannot afford the other fruits and your efforts will bring other fruits to their menu. There has been "apple revolution" due to high apple production in Himachal and Jammu and Kashmir. But similarly revolution needs to take place in peaches, pears, plums and apricots, walnuts and almonds in suitable areas.

Perishable Fruit : Strawberries and other similar kinds of fruits are available to the big cities for only a very limited period. These are also perishable fruits, but their increased production has a long way to go and the horticulture students who will be the young entrepreneurs of tomorrow can do a lot in this respect also.

I also recommend a "Papaya Revolution" for the foothills and plain area whereas in urban and suburban and rural areas, you should stimulate production by all of dwarf fast grows honey dew or similar variety of papaya, which starting fruiting within one year. Get papaya planted in millions --good for other purposes.

Organic Recycling : Trees for planting need organics. Build up more compost and organic matter for the initiation of healthy growth of fruit and other tree. Use of all waste including huge fall of trees leaves for leafmold and other organics is becoming increasingly important.

Summing up : Horticulture, I believe is both an art and science and many of you have learnt this and you can adopt it to the large-scale nursery plantations and also for agro-forestry operations. You can help develop large-scale plantations in the foothills and other suitable places for other species including subabul, eucalyptus, sisoo, kiker and various other types of trees which are needed for fuel, fodder, construction and paper needs of the country. I would recommend that the project preparation is an important link from theory to an action oriented programme. Prepare a suitable project. Be bold, ready to work hard and you will not find deficiency in land or even in investment. Banking institutions, many non-governmental and governmental organisations will be ready to help you in promoting horticulture in the Himalayan areas. In this process, you will serve the Himalayas. You will bring money to the Himalayan people. You will promote employment and you will bring prosperity to the area.

5

Eco-system of Kandi Area in the Outer Himalayan of Jammu (J & K State)

MRS. KIRAN CHADHA

Introduction

The water scarce land in Jammu is known as the Kandi area. This land, stretches between 32° 20′ to 32° 55 N′ and 74° 10′ to 75° 50′ E and falls mostly to the West of Ravi river. The width of the area ranges from 10 to 40 km and ranges in attitude from >300m to <600m above the mean sea level. The whole terrain is rugged, mature, uneven and represents a look of semi-arid land. In length it extends over a distance of 250 km from Ravi to Manawar Tawi. The important streams and rivers include the Chenab, Ujh, Basantar, Tawi, etc. Topographically speaking the area is undulatory, barren, broken and being intersected by dry nallahs, Khads, ravines, gullys, etc. There are gigantic escarpments separated by broad longitudinal strike valleys. It represents a view of barren landscape.

Geographic Set Up

The Kandi area has a typical sub-tropical climate with extremes of temperature with summers very hot and winters dry and cold. The summer maximum is 46°C and winter maximum being 22°C. The month of December records further fall in day and night temperatures. Occurrence of mist and fog locally known as DHUND and KOHRA respectively is common in the morning hours of January. The monsoon rains are noticed between July and September. The atmosphere during the period is quite humid with frequent lashes of torrential rains. The average annual rainfall is less than 100 cms.

The terrain has a thick undergrowth of bushes and scrubs. The dominant species include "Neem", "Pipal", "Tun", "Kikar", "Babul", "Khair", thorny bushes, evergreen Shrubs, Climbers and the tall grass locally called as "Khar". Most of these species are broad leaved deciduous type. The soils are alluvial, sandy, loamy and admixture of sand and clay. Geographically, the Kandi terrain covers the shivalik super group. The rocks are sandy chiefly composed of sandstones, grits, conglomerates, pebbles, gravels, silt and clay. The rocks are of fluviatile origin derived from the sub-aerial erosion of the mountains. The debris thus produced by erosion was brought down by numerous rivers and the material was finally deposited at the foot of the Himalaya. The Siwaliks have been involved in the later phases of Himalayan Orogeny, for we find them often folded, faulted, overthrust and lying at steep angles against other formations.

The water table in the Kandi area is at a depth which varies from 100m or above. The soil has low moisture retaining capacity. The problem of regreening the Siwaliks is both economic and ecological. Lakhs of people living in the Kandi area have no adequate source of livelihood—no crops, no natural resources, no industry and no raw materials. They are purely dependent upon their livestock or jungles. To meet their daily requirement for fuel, fodder, etc. they have cleared the jungles. With the disappearance of forests, the ecological set up has been disturbed and now there is no water, no fuel and no fodder.

According to Maj. General Goverdhan Singh Jamwal, AVSM (Retd.), who started campaign "Save the Shiwaliks" and initiated a padyatra along with his other members of the local "Paryavaran Sanstha" on 6th Feb. 1988 stated, "without trees, water and livestock, the Shiwaliks have been reduced to a mountain desert unable to sustain such a high density of population. . . ." The area is full of ravines which carry flood waters during the monsoons and the soil is exposed in the form of sandy beds by ephermal streams. The complex geological history of the Himalayas and the tectonic stresses have contributed to aggravating to the earths' dynamic process of weathering, erosion and mass wastage, which effect and modify the natural environment.

Problems : Ecological and Environmental

The unscientific cultivation along the slopes, uncontrolled grazing by increasing number of livestock, excessive deforestation for meeting local firewood and fodder requirement as well as timber needs of the buildings and industries. This has led to sharp degradation in the eco-system of the outer Himalayan regions. There is an urgent need for tackling the eco-rehabilitation problems. Basically the objectives should be to stop erosion, to increase productivity of land for increasing the living standards of the inhabitants. On the one hand, these problems are independent but on the other, they are interdependent and have influence over the economic well being of the masses and the local eco-system.

Irrigation

The Kandi area of Jammu experiences the extremes of whether and climate. The scorching heat in summer and their erratic behaviour of monsoons force the farmers to irrigate their crop, wherever irrigation facilities are available. Kandi is in the grip of a water famine. What to speak of water for irrigation, there is no drinking water. Human and animal life is in distress. While the old ponds are lying dry or in disuse and no effort has been made to construct or dug new ponds. The tube wells are out of order and at many places not working because of very low water table. The fall in the water table can be ascribed to ruthless and unplanned felling of trees and deforestation. Long queues of village women waiting for water supply tanks for hours is a very pathetic scene in Kandi villages around Kathua, Samba, Akhnoor, Amb Gharota, etc. A Bakarwal has to roam about 15 to 20 km daily in search of water for his livestock. It is not paradoxical that six perennial streams including two major rivers *i.e.*, the Chenab and the Ravi, are flowing through this Kandi belt yet inhabitants are facing a water famine. It is not impossible that the water resources cannot be tapped for human consumption or irrigation. What is lacking is the personnel, professional and political effort by the people.

The nallahs and streams are not perennial and can be dried off quickly. In an area to the North of Dhar-Udhampur road, Springs' water is used for irrigating the lands. In this belt, besides dry crops, such as Maize, Barley, etc., there is cultivation of paddy in some isolated areas. In some longitudinal valleys or KHADS the rice cultivation has replaced wheat, otherwise the area on the whole is agriculturally poor.

To develop agriculture, the waters of the Ravi, Chenab, Basanter, Tawi, Ujh can be used to irrigate lands. The area is inhabited by very hard working people, Gujjars and Bakarwals, who have migrated to this belt since long back.

According to Dr. B.L. Chaku the increase in population and also in their livestock has exerted a great pressure on the land. There has been shrinking in the pastures and grazing lands. It has enhanced the soil erosion, sheet floods and disturbed the ecological balance. The frequent droughts have further aggravated the situation. Keeping in view the principle of demand and supply with respect to fodder, there is urgent need to freeze the present livestock population. Due to social and religious sentiments involved drastic reduction in livestock population is impossible. However, the Gujjars and Bakarwals should be motivated to keep few improved species of animals in place of a herd of useless and uncommercial animals to draw maximum benefits by providing adequate fodder, which at present is being shared by poor and uneconomical animals.

Though reduction in livestock population will relieve pressure on the grazing land, however, the ultimate solution to the problem would be the improvement of forage resources and scientific utilization of herbage available from these resources. The measures to be adopted for increasing forage supply would call for diverting some pressure from grazing areas to the agriculture land, waste-land by way of popularising production of fodder crops and grasses on farm boundaries, grasslands, orchards and on the cultivable land where such land can be spared.

Cultivation of superior fodder will not only increase supply of adequate and nutritions fodder but also enable the regeneration of pastures and grazing areas by relieving pressure on these areas. The widespread menace of weeds like "Lantana", "Xanthium", etc., should be checked in the Siwalik hills which are responsible for destabilising the ecological balance.

Horticulture

The Kandi belt, a potentially horticultural belt is undeveloped yet for horticulture. A rationale land use pattern also favours big horticultural enterprises strongly for Kandi areas which is most unsuitable for food grains production. The World Bank experts are also convinced of this. Horticulture in the Kandi belt without irrigation is just unthinkable. However, mangoes grow in abundance. The quantity is good but the quality needs improvement. The other fruits include Ber, Plum,

Papaya, guava and grapes. The mango crop takes about four months to mature. For a steady expansion of horticulture, the grower must provide precise inputs and be vigilent against pests and disease. In fact, exact scheduling is essential as most of the fruits and flower crops are highly perishable.

The Problem of Landless People

In the Kandi area, the economic development has been very slow. One of the numerous factors is the limited land-holdings. In certain cases, the land holdings are less than half an acre. A large number of people of the weaker sections of the society are landless. A large part of government land is lying vacant. An earnest effort in this direction is to offer the land to landless and ex-servicemen, who are strong and sturdy and most suitable to develop this area. It will achieve two objects firstly, the unused land will be brought under agriculture or horticulture and secondly, the economic condition of the inhabitants will improve.

Social Forestry

Keeping in view the general degradation in the ecology and ever increasing gap between supply and demand of fuel and fodder, the International Development Agency and USAID jointly came forward to finance Social Forestry (Umbrella) Project for a number of States in India. Jammu and Kashmir State is one of the States where this project has been launched and aims at afforesting of land. Social forestry has essentially to be fitted in the regional commitment for afforesting available Kandi lands. Simultaneously, social forestry has to increase productivity of available land resources to satisfy the basic needs of the people. Under this programme, really healthy and well grown plants have to be distributed so that failures do not occur in the field. Except a few stagnated eucalyptus saplings no rewarding work has been done in the Siwalik Kandi. Even bamboo plantation has not been taken up seriously. Eventually the social forestry project executives have yet to evolve a planting technique for Kandi which ensures moisture conservation, choice of species, depth of planting and tending operation. It is a problem of dryland afforestation. Hence the need for a new and special technique. While the soil conservation and moisture conservation techniques can help establishing plants in Kandi, sustainable development of plantation is possible only with irrigation.

Kab Bamboo

The Siwalik Kandi belt is potentially a bamboo belt. Without irrigation even bamboo cannot thrive and develop into viable economic plantations although it may contribute towards the development of bio-mass for environmental and visual impacts. The local Paryavaran Sanstha members while on padyatra also observed the presence in abundance of another important tree called "Siris" which could be employed for timber and fodder, The natural bamboo forests of Kathua and Jasrota area can be developed.

Recommendations for Kandi Area Development

(*a*) The fauna and flora should be preserved for posterity and for scientific research.

(*b*) The crop and animal productivity of the Kandi area should be improved. Since conditions very greatly within the region because of variations in soil, climate, topography, etc., it is, therefore, necessary to provide the farmers latest technological know how and methods of agriculture for improved and suitable food crops fodder and fruits.

(*c*) The pasture lands should be developed and suitable leguminous plants and fodder grasses should be introduced.

(*d*) The overgrazing leads to degradation of the grass cover and the erosion of the precious top soil. It is, therefore, recommended that there should be a detailed study of situation in respect of domestic animals, with the object of determining the most suitable breeds and populations which can be maintained on the basis of the carrying capacity of the Kandi area.

(*e*) Recognizing the importance of vegitation cover and sound landuse practices, to soil and water conservation, it is recommended that co-ordination among the irrigation, soil conservation and forest departments is necessary. This co-ordination would ensure optimum social and economic returns to the local population without impairment to the environment.

(*f*) The evolving of sound management practices should form the core of social forestry schemes. The suitable species of trees should be raised which can meet the local needs of timber and fuelwood should be intensified.

(*g*) There should be a total ban on felling of trees and plantation drive should be on war footing.

(*h*) To make an arrangement for adequate supply of drinking water, the old tanks, ponds and springs should be properly looked after, repaired and improved.

(*i*) Landless people should be allotted land to raise the plantation.

(*j*) Lantana should be replaced either by turning it into charcoal or replaced by bamboo or other local species.

(*k*) Biogas Smokeless Chullhas, Solar Cookers, wind mills and other non-conventional sources of energy should be introduced in the Kandi area.

(*l*) To prevent water run of and use it for protective irrigation, small reservoirs, specially designed to suit the local topography, soil and climate should be constructed.

(*m*) The economic condition of the people should be improved by offering them loans, credits and financial assistance so that they may develop their ancestrol professions such as shoe making, brick making, leather tanning, etc. The construction of roads, opening of schools, supply of cheap electricity for developing collage industries subsidies for poultry farming, bee keeping, sheep breeding, etc., will definitely improve their socio-economic structure.

REFERENCES

Chadha, Mrs. Kiran. 1987. Tourism and its impact on environment in the Himachal Himalaya., Himachal Himalaya : Ecologoy and Environment Ed. : S.K. Chadha. Pub. To-Day and To-Morrows' Printers and Publishers, New Delhi, pp. 121.

Chadha, S.K. 1987. Himachal Himalaya : Ecology and Environment (Edited) S.K. Chadha, Pub : To-Day and To-Morrows' Printers and Publishers, New Delhi.

Chadha, S.K. 1988. Himalayas : Ecology and Environment Pub : Mittal Publications, Delhi.

Chaku, B.L. 1985. Geographical Problems and Development of Kandi Area in Jammu Division, in Geography of Jammu and Kashmir (Some-Aspects), Compiled by Majid Husain, et. al., Pub : Ariana Pub. House, New Delhi, pp. 289-293.

"SAVE SHIWALIK"—A REPORT

Jamwal, Goverdhan Singh 1988. Daily Excelsior, May 19, 1988. Vol. XXIV, No. 136, pp. 4-7.

Raina, A.N. 1971. Geography of Jammu and Kashmir, Pub : National Book Trust, New Delhi.

6

The Environmental Set up of Ladakh Region of J & K State, in Western Himalaya

MRS. KAMAL SAHNI

Introduction

The State of Jammu and Kashmir comprises of three divisions *i.e.*, Ladakh, Jammu and Kashmir. Ladakh constitutes the northern most administrative unit of the State. It is a vast arid table land located at a height of 5,000 metres. This hostile, barren treeless zone is intersected by high vertical cliffs of black grey and brown sand, the stratification of which is arranged in such a manner that they look superimposed upon each other, having their vertical faces to the streams or river. At certain places, there are broad and flat valleys and some of them are laden with fruit and its lush meadows are a heaven for the animal wealth. The region is popularly designated as the "Root of the World" and people live at an altitude ranging from 2,800m to 5,000m above the mean sea level.

The mightly river Indus gushes through diverse types of rocks and flows majestically. On the left of the Indus is the Zanskar range, while on the right is the Ladakh range. Indus, though flowing through gorge has numerous alluvial fans and river terraces. The town of Leh is situated on an alluvial fan. To the East of Leh, on either bank of the Indus is an alluvial plain stretching over a distance of 30km. This is the most fertile levelled land sprinkled with several rural settlement of the Ladakh region. The flow of the Indus fluctuates, being highest during summer and lowest during winter. There are floods in the river during summer and droughts during winter. The scanty rains and arid conditions are not favourable to promote agricultural activities and therefore, cultivation of crops is confined to the terraces of the river Indus. The settlement is sparsely distributed and the density is low.

In the northern parts of Ladakh are seen the Karakoram mountains, which radiate to South-eastern direction from the Pamir Knot. Some of the highest peaks of the Himalaya *e.g.*, K_2 or Mt. Godwin Austen (8,610m) are situated within the Karakoram Range. This mountain range has also abode of glaciers, *e.g.*, the Saichen, which has a lengh of about 72kms. The other valley glaciers are Baltore, Hispar, Rimo, Baifa, Batuna, etc.

Population

The spatial distribution of population depends upon several factors such as terrain, geographic location, physiography and relief, topography, climate and natural resources. The Ladakh is a negative area in the sense that none of the natural factors support a good population. The natural and environmental set up is such that the population is sparse and thin. The population distribution, according to 1981 census, is as under :

TABLE 6.1

Showing District-wise Population Distribution

	District	Population	Percentage	Total
1.	Leh	67,733	1.14	
2.	Kargil	64,566	1.08	
	Ladakh Division	132,299	—	2.22

Kargil and Leh have a density of 5 and 2 persons per sq. km. respectively and are the lowest populated districts of the Jammu and Kashmir State. The number of households in kargil is 0.88 per sq. km, while in Leh, it is 0.33 per sq. km. whereas the average of the State is 9.27 per sq. kms.

The age sex, and size are the chief components of population and their literacy rate, religions and occupational characteristics are the vital components of social and economic development. In Ladakh the sex ratio is at par to the State average, being 879, *i.e.*, 879 females per thousand of males. Literacy in rural area of Ladakh is very low and is even lower in the case of females. Religion is an important feature of population and in Ladakh the composition is as given in Table 8.2.

TABLE 6.2

Showing Religions Composition of Ladakh

Muslims	Hindus	Sikhs	Buddhist	Others
44.66%	0.01%	—	51.82%	3.51%

The district of Leh has a high concentration of Buddhists while Kargil district is dominated by Muslims. In Ladakh only 0.97 per cent of the urban population resides and the areas of urban population are Kargil and Leh. About 45 per cent of the total population of Kargil and Leh towns is dependent on primary activities, about 16 per cent work in transport, 30 per cent in services, 6 per cent construction and only 3 per cent is dependent on manufacturing.

Climate

The climate of Ladakh and Zanskar is very cold, arid and dry. In winters, temperatures are extremely low in Rupsho, the higher plateaus above Tansktse, Rong and Zanskar, while Central Ladakh and Nubra Valley possess a comparatively mild climate. During July Leh has 25°C and 7°C, the mean maximum and the mean minimum respectively. During this month, the diurnal range of temperature is significantly high, being about 18°C. In January the mean minimum temperature remains below freezing point, reading as low as —16°C at Leh and —40°C at Dras. During the month of September to May, the night temperature at Leh remains below freezing point. The skies are often clear and insolation very intense. The average annual rainfall is only 20 cm at Leh, most of which is received in the months of March, August and January. The precipitation is shown in Tables 8.3, 8.4 and 8.5.

As a result of the peculiar precipitation, rainfall and cold climate, Ladakh has unusual landscape of stunted and gnarled shrubs of small trees of "Juniperus" with stunted, tufted patches of "Stachys tibetica" and

TABLE 8.3

Showing Winter Precipitation in Some of Districts of Ladakh

District or Station	*Lat.*	*Long.*	*Height a.m.s.l. in m.*	*Winter Precipitation in m.m. (Nov. to March)*	*%age of annual Precipitation*	*Annual Precipitation in m.m.*
Dras	34° 26′	75° 46′	3,066	312.7	41	756.9
Leh	34° 09′	77° 44′	3,514	28.4	24.6	115.0

TABLE 8.4

Showing Summer Monsoon Rainfall in Some Districts of Ladakh

District/ Station	*Lat.*	*Long.*	*Height a.m.s.l. in metres*	*Monsoon Rainfall in m.m. (June-Sept.)*	*Percentage of the Annual*
Leh	34° 09′	77° 34′	3,514	51.7	45
Kargil	34° 34′	76° 08′	2,682	33.7	11
Dras	34° 26′	75° 46′	3,066	65.5	9

TABLE 8.5
Showing Post-Monsoon Precipitation in some Stations of Ladakh

Station	*Post Monsoon rainfall (Oct-Nov.)*	*Percentage of the Annual*	*Annual Rainfall in m.m.*
Dras	40.7	5	756.9
Kargil	9.1	3	306.5
Leh	10.0	9	115.0

conspicuous dense cushion like shrubs of "Caragana versicolor", "Acantholimon lycopopiodes" and "Thylacospermum Caespitosum". Some of the other plants here are the mat-farming "Hippophae", Carex" spp., etc., "Microula tibetica" is an unusual herb belonging to the family Boranginaceae, with outspread leaves enclosing a dense cluster of flowers at the centre.

Geology

The Ladakh region of the Jammu and Kashmir State carries one of the finest developed sequence of rocks right from Archaen Era to Sub-Recent age.

Pre-Cambrian rocks are developed in Ladakh and Zanskar and are designated as Salkhala Formation. The rocks are constituted of states, phyltites, quartzites, mica schists, carbonaceons and graphitic schists, crystalline limestones, dolomites and biotite Gneisses. They are highly folded and compressed. The rocks have been involved in the movements which brought the Himalaya mountains into being. The rocks are often found thrust over the rocks of Permo-Carboniferous or later ages.

The Salkhalas and the Gneisses are traversed by later igneous intursives including gabbro, pyroxenite, dolerite, hornblende-granite, tourmaline granite and pegmatite. The panjal volcanics of upper carboniferous age are excellently developed to the West of the Zanskar range upto Hazara, in Ladakh, etc. The lower part of the formation is known as "Agglomeratic States" while the upper part is designated as "Panjal Traps". The volcanic agglomerates are gritty or greywacke like and grade frequently into states. They contain angular fragments of quartz, feldspar, quartz-porphyry, granite, limestone, devitrified glass, etc. At places, these rocks carry various genera of brachiopad fossils, bryozoans and lamellibranches.

The Panjal Traps consists of bedded flows of green and purple and dark colours. The laras are sometimes amygdaloidal or porphyritic. The flows of lava vary in thickness, andestic to basaltic in composition. The rocks are unfossiliferous.

The Triassic rocks of Ladakh are composed of limestones known as "Megalodon Limestone" and are overlain by Jurassic rocks, designated as "Spiti shales". These contain fossils of ammonites and Belemnites. Some fossils of lamellibranches and brachiopods are also present in them.

Cretaceous rocks in Ladakh are designated as "Indus Flysch" and "Dras Volcanics". Cretaceous limestones with "Gryphaea vesiculasa" and other fossils have been encountered in Rupshu and along the Leh-yarkand road in sinkiang. Cretaceous deposits are known in the upper Hunza valley and in Eastern Karakoram and also along the Indus valley North of the Himalaya. The Indus river North of the Himalaya flows along the flysch zone which marks a major tectonic zone which Gansser (1964) calls the "INDUS SUTURE".

The volcanics accompanying the flysch zone are named after Dras. Here both lower and upper cretaceous rocks enclosing "Orbitolina" and other fossils are found. The volcanics consist of thin bedded tuffs and ash beds of purple and greenish colours with layers of agglomerate, Shale, Cherty Jasper and lava flows. The lavas are basaltic and andestic but are generally chloritised and epidotised. The intrusives are partly contemporaneous and partly younger and are composed of norites, peridotites, pyroxenites, hypersthene-diorites. Granites also occur, accompanied by parphyries and dolerite dykes and sills. The granites are of post cretaceous age.

The new allurium of sub-recent age has been formed along the Indus and it comprises fine suit, loam, clay and fine grained sand with fresh molluscs and other invertebrate organisms. The town of Leh is situated on the alluvial deposits.

The erosional agencies have played a vital role in producing diverse types of geomorphic features in and around Ladakh.

Soil

Soils are directly or indirectly formed by the weathering of rocks, decayed organic matter and moisture. The soils in Ladakh region of the Jammu and Kashmir State, are cryoborolls (Mountain meadow Soils). Such soils are found at high elevations, with dry and cold climate and scanty vegetation.

The soils are shallow to moderately deep and immature. They suffer from moisture deficiency resulting from prolonged drought, wind erosion and snow action. By and large, they range in texture and structure because of admixtures of disintegrated rock fragments, colluvial materials and screes. Only in the bottoms of valleys and in law lying terraces, sandy soils of varying depth are found. In some places, they are gravelly sandy loams.

Lithic Entisols (Skeletal Soils) are found to the North-eastern part of Ladakh. The soils are very shallow, overlying the weathered mantle of parent rocks. Owing to deficient precipitation, poor vegetative cover and the existence of condition unfavourable to intensive weathering, either chemical or mechanical, soil development is limited. If there is any soil cover, it has been badly affected by wind erosion. In most of the areas, the base rock is exposed and is ecologically a havoc.

In general acidity in different proportions is associated with most of the soils. Another problem is the high intensity of water erosion. Owing to steep slopes, erratic land use, indiscriminate grazing, and the lopping and felling of trees at random, the problem of soil erosion has become acute and an environmental liability of the region.

The soils are utilised for the cultivation of vegetables, grim small millets, fodder and orchards. The planting of medicinal plants and herbs alongwith the existing ones, with proper management should be encouraged to maintain the eco-system in this cold desert.

The soil erosion can be checked by maintenance of an affective vegetation cover, contour ploughing, crop rotation, terracing, compositing, creation of wind breakers, fencing, pipe drainage to prevent gullying, damming of gullies, etc.

Occupation

The State of Jammu and Kashmir is essentially an agrarian in which most of the people are engaged in agriculture, horticulture, apiculture. The export from the State includes fruits, wood and wood products, gums and resins, dry fruits, hides and skins, wool and wool products, papier mache, turpentine oil, silk, saffron, seeds, honey, marble, tea, fur goods and embroidery goods. Despite many incentives, the process of export is slow and the material exported is mostly raw. A pragmatic policy is to be followed to enhance the export of processed goods.

Ladakh being a cold region is connected with the rest of the State by road and air. The region is rugged and sparsely populated. People are engaged either in agriculture or dry fruit export. During winter, most of the population depends upon indoor work including knitting, handicrafts and embroidery. People also depend for their living on rearing goats, yak and fur sheep. The goats bear a special wool called PASHMINA—a natural incubator for these animals. This fibre is the source of one of the major cottage industries in Ladakh, and is used to produce Jamwar and Pashmina Shawls.

Environmental Problems

Although most of the environmental problems are similar to those of Lahul and Spiti in Himachal Pradesh yet some of the specific problems of Ladakh need discussion. This region has a short lived summer and long freezing winters. There may be snowfall from September to April. The habitations is thin and concentrated either in Leh or in Kargil areas. The winter precipitation is in the form of snow. During heavy snowfall the mercury goes much deeper to minus freezing point. The snow surface becomes hard, breath turns into iciles and face skin peels off due to tanning. At times, the snow storms, cloud bursts and avalanches cause a great loss of life and property. Before the commencement of winter, the people set out to collect fuel, food and fodder. They also collect dung cakes from pastures and uproot every twig and shrub from the valleys and mountain slopes. It is their prized possession and a status symbol which is proudly stacked on the roof tops. As a result, there has been great deforestation and the green vegetative cover is fast disappearing.

Apart from the civil population, the military and para military forces also make a heavy drain on the forests to meet their fuel and energy requirements. The strong solar radiation is responsible for the rather high maximum temperatures at Leh. The highest temperature recorded at Kargil (26 82m) is 43.3°C. The solar energy can be harnessed to make use of non-conventional sources of energy in this region.

In this mountain desert, there is an abundent wind power. The howling winds, with a velocity of more than 50 km. per hour sweep the valleys. The wind energy can also be harnessed by installing aero-generators and wind mills for power generation. Recently, some work has been carried out to instal solar heaters. The results are very encouraging.

The denuded, rugged and treeless tract requires afforestation along river beds and villages peripheries. Thus, the natural forests are remarkably efficient eco-systems having both protective and productive functions with an important role in the economic development of the region. Evidently, not only further destruction of the forest is to be checked but also an additional land has to be allotted for the creation of large scale biomass resources of suitable fast growing species for production and protection forestry. The fuel wood consumption, urbanization and the energy crisis would continue to have a vital role in rural life and people will have to depend upon biomass energy. This aspect must be given due importance while making management plans.

The barren regions include all those lands where the run off is very high, forest cover is minimal and water scarcity in summer months has become a regular feature. Therefore, the strategy should be to restore eco-balance of degraded lands, improvement of the environment and to meet the requirement of fodder and fuel.

The cattle population is surviving on grazing, by moving from one to another, particularly to high attitude pasture lands in summer and rainy season. Obviously, the barren and wastelands are the main stay for grazing for the semi-nomadic groups. It is, therefore, necessary to improve the pasture lands. The introduction of new cropping pattern will have to be identificd judiciously so that the sustained productive use of such barren land is ensured. There is also a need to give high priority to the plantation of bamboos and suitable tree crops along the boundary of fields as well as in the barren lands for stall feeding of exotic cattle.

In the areas identified for development where wild species fruits are a part of the habitat, these should be upgraded and improved by quality fruits varieties of the same species. Particular stress be laid to improve wild ber, olive, peach, etc. The scheme will least disturb the natural associations in the selected habitat. The land use should be carefully improved and changed on sound scientific basis so that the biotic equilibrium of the natural speciation is not irreparably altered.

Causes of Wasteland Formation

The accelerating speed of soil erosion and soil degradation culminating into the formation of wasteland in Ladakh may be attributed to the following :

(*a*) Improper and unscientific approach in land-use management. Careless and improper farming methods such as ploughing up and down hills or removing all vegetation on tracts of lands exposing the soil to extensive erosion.

(*b*) Increasing deforestation of large area on steep slopes to cater the needs of local inhabitants for fuel-wood *vis-a-vis* clear cutting of timber for commercial gains.

(*c*) Over grazing of pastures and range lands interferring with the natural succession of plants and killing grasses and forest that naturally hold the soil in place. In Ladakh, more than 70 per cent of the fodder requirements are met with from forests through grazing, grass cutting and lopping of leaf fodder.

(*d*) Construction of structures in forests, heavy road construction in the forestlands particularly in the ecologically fragile areas. Indiscriminate use of forested and fertile lands for house construction, etc. All this leads to depletion of food, fuel, fodder, fruits, etc.

(*e*) Most of the people in the region don't have sound financial status to develop the arid and wasteland on scientific basis. In fact, wastelands and poverty co-exist together.

The Impact of Tourism on Environment

Ladakh is a bewildering creation of nature, unique in land architecture, mountain scenery, meadows, precipitous gorges, valleys, glaciers, springs, etc. They are rich in floral and faunal wealth and are tempting for climatology of recreation. Because of inaccessibility and having difficult geographic terrain, most of it has escaped human erosion. It is a region of ecological pursuits where pastoralism, hill terracing, fishing and hunting have dominated the scene. Still the region majestically stand in its primeval glory, simplicity and uncommercialized character.

During the recent years the number of Indian as well as foreign visitors has increased several folds. Closely tied up with the problem of tourism is the issue of protecting and conserving the environmental resources which the tourists in their wanton and holiday making moods erode or degrade directly or indirectly and often unintentionally. This is important both for tourism and environmental, as the quality of "tourism products" depends upon a high quality of natural environment.

Conserving the Wild Life

All the mammals of Ladakh are provided with long and thick hairy covering at least in winter to cope with the severe cold. Hibernation undergone by some animals is an effective means of tiding over the winter. The animals capable of hibernating store up excessive fat as reserve food. Several animals sresort to migration to lower altitudes during winter. The rapid urbanization, indiscriminate poaching of wild life for commercial purposes, illegal trade in protected animals, trade in animals used for laboratory experiments and disease are major causes of extinction of wild life.

For effective and successful conservation of life, it is necessary to check the deforestation which provides ideal habitates. The poachers should be brought to the book and be imparted rigorous imprisonment. Spray of pesticides and insecticides be stopped and replaced by other pest control devices. Breeding and conservation of rare species of wild animals be given priority over other ecological and eco-system programmes. The wooded areas rich in wild life stock should be declared as bioreserves.

There is also a good possibility of discovering new taxa in the Ladakh region because of high mountains and rivers acting as barriers to natural distribution.

Some Recommendations

The Ladakh region plays a vital role in the social, cultural and economic life of the people of the State. The Ladakh Himalaya influence the climate, contain the sources, the catchments and watersheds of major river systems of the State. They abound in forests, plant and mineral wealth, their potential for yielding mineral resources, geothermal power and energy is appreciable. The Ladakh Himalaya are also the home of sizeable local populations of diverse cultural heritage and richness, and their attraction as centres of tourists resort. Keeping in view the rapid development of the region, it is necessary to incorporate environmental and ecological considerations to ensure the ecological balance.

Keeping in view the rich potentials of the plants, a detailed survey should be conducted to study the less known plants, herbs, medicinal plants, etc. Steps should be taken to preserve to rich animal population. Measures be taken for the improvement of both crop and animal productivity of the region while planning agricultural development in an integrated manner to preserve the environment. It is recommended that studies be undertaken to know the existing cropping systems to further understand the rate of run off, soil erosion, etc. At the same time, agronomic experiments should be initiated with the object of identifying cropping systems and rotations which do the least damage to the environments while bringing to the farmer optimum returns from his land. A well thoughtout programme be initiated for the expansion of horticulture in this hilly region.

In the Ladakh region, where pashmina goats are reared, but where because of an of acute temporary shortage of fodders, high mortality occurs it is recommended that food banks be established to deal with the situations which otherwise leads to loss of valuable animals. The effects of various development activities on the eco-systems in the Ladakh region need to be monitored continually. To save the tourists from snow-storms, avalanches and landslides, there is a dire need to collect meterological data regularly through observatories. For this purpose, automated and telemetric stations should also be set up. The data gathered by these stations can be supplemented with remote sensing, using satellite imagery.

An important segment of the strategy of development of the region would lie in the multidimensional use of land through a suitable mix of animal husbandry, agricultural and forestry practices, supplemented with subsidiary occupations, such as floriculture, apiculture and sericulture.

Since Ladakh is famous throughout the country for its thermal springs, it is therefore, useful to tap these for geo-thermal power. A considerable thought has to be bestowed on the very parameters of settlement planning in this region. To bring about the socio-economic development of this hilly terrain, it is necessary to evolve appropriate planning techniques, standards and norms for the planning of human settlements.

REFERENCES

Bose, S.C. 1972. Geography of Himalaya, Pub : NBT, Government of India, New Delhi.

Census of India 1981. Series 8, Jammu and Kashmir, Government of India, New Delhi.

Chadha, S.K. 1976. A Text Book of Geology, Pub : Kapoor Bros., Srinagar.

——, 1987. Himachal Himalaya : Ecology and Environment (Ed.), Pub : To-Day and To-Morrows' Printers and Publishers, New Delhi.

Chib, S.S. 1977. This beautiful India : Jammu and Kashmir Pub : Light and Life Publishers, Jammu.

Hussain, Majid. 1987. Geography of Jammu and Kashmir State, Rajesh Publications, New Delhi.

Lawrence, W.R. 1967. The Valley of Kashmir, Pub : Jay Kay Book House, Jammu.

Malkania, N.P., Malkania Uma and Tondon, J.P. 1988. Design for and conservation of Environment in Himalaya in S.K. Chadha (Ed.), Himalayas : Ecology and Environment, Pub : Mittal Publishers, New Delhi.

Malkania, N.P. and Malkania, Uma 1987. Man and Environment in Himachal Pradesh in S.K. Chadha (Ed.), Himachal Himalaya : Ecology and Environment., Pub : To-Day and To-Morrows' Printers and Publishers, New Delhi.

Raina, A.N. 1971. Geography of Jammu and Kashmir, Pub : NBT, Government of India, New Delhi.

Sahni, Kamal. 1987. Rich Potential of Thermal Springs in Himachal Pradesh in S.K. Chadha (Ed.), Himachal Himalaya : Ecology and Environment Pub : To-Day and To-Morrows' Printers and Publishers, New Delhi, pp. 55-60.

Wadia, D.N. 1919. Geology of India, Pub : Tata Mc Graw Hill Publisher Co., New Delhi.

7

Ecological Planning of Uttar Pradesh Himalaya

KATAR SINGH

Introduction

The Uttar Pradesh hills extend over an area of about 51,100 sq.km. which is about 17 per cent of the total geographical area of the State. Agro-climatically, the Uttar Pradesh hills represent a distinct region called the "hill region". According to the 1971 Population Census, the total population of the hill region was about 3.8 million which was about 4 per cent of the total population of the State. Administratively, the U.P. hills are divided into two divisions, namely, Kumaon and Garhwal, and into eight districts, namely, Almora, Nainital and Pithoragarh falling in the Kumaon division and Chamoli, Dehra-Dun, Garhwal, Tehri and Uttar Kashi falling in the Garhwal division.

The hill region occupies an important place in the economy of the State. It is an important source of supply of temperate fruits and off-season vegetables in the State. Its forest resources provide us with timber, fuel wood, fodder, medicinal herbs, resin and a number of other minor forest products. Its water resources are used for generating hydro-electricity and providing irrigation and drinking water. Its high mountains provide a natural protection against the icy cold winds from Siberia in winters and a natural defence against any enemy attack from the North.

The hill region is also an important source of supply of manpower to the rest of the State and even to some other States in the Country. The people hailing from the U.P. hills are known for their simple mindedness, honesty, sincerity and loyality. It is because of these personal traits that there is great demand for the services of the people from the U.P. hills in a number of public and private concerns.

Historically, the region has been famous for its calm and quiet environment conducive to the pursuit of *Yoga*. The holy Ganga river and its tributories have all their origins in this region and the legend has it that many a Hindu god and goddess had their abode in this region. And that is why the region attracts lots of pilgrims from all over the country every year. Besides, there are quite a few places in the region which because of their cool and pleasant weather in summer attract a lot of Indian and foreign tourists.

Thus, in a nutshell, we could say that the U.P. hills occupy an important place in the economy of the State. But in spite of this, it is surprising to note that due attention has not been given to promote economic development in this region. As a consequence, the region continues to be economically depressed and has been by-passed by the recent wave of Green Revolution. There is no semblance of a comprehensive strategy for economic development of this region. What exists in the name of development plan for the region is no more than an uncoordinated assortment of selected activities falling under the jurisdiction of separate government departments and agencies. What is needed for promoting an all-round development of the region is a comprehensive strategy based on integrated area approach. This study is intended as a modest attempt to outline the guiding principles and major elements of such a strategy. The analysis mainly focuses on agricultural development defined to include substantial increases in agricultural output, income and employment in the region.

Before the new strategy is outlined, a brief description of the salient characteristics of the economy of the region and a critical review of the development strategies followed in the region in the past seem in order. The former is intended to give an idea of the existing state of the system; its problems and potentials and the latter to serve as a back ground against which to evaluate the appropriateness of the new strategy.

2. Major Characteristics of Hills Economy

Agricultural is the dominant sector of the economy of the region with about 80 per cent of the people dependent on it for their livelihood. It is characterized by predominance of marginal and small farms, fragmented and scattered land holdings, lack of irrigation facilities, prevalence of traditional methods and practices of crop and livestock production and low crop and livestock productivity. According to the Agricultural Census, 1970, about 68 per cent of the total number of operational land holdings in the region were less than one hectare in size and about 94 per cent less than three hectares. The average size of land holdings was 1.07 hectares which is slightly smaller than the State average. The land holding of the average farm was scattered over about 8 locations and on the average there were three fields per location each measuring about. 03 hectare (5)*. Due to non-adoption of proper conservation measures, land and water resources of the region have been continuously deteriorating over time. Under the increasing pressure of human and livestock populations on land, steep slopes which are inherently unfit for arable agricultural are being brought under cultivation. The community lands which account for about one-third of the total geographical area of the region and comprise panchayati and civil forests, culturable waste, unculturable waste grazing lands and land under miscellaneous tree crops are badly eroded and completely denuded of their vegetation as a result of uncontrolled over-grazing and indiscriminate lopping. In the absence of appropriate institutional structure for their management, these lands are the victims of what is known in the literature of welfare economics as 'tragedy of commons', (2). The productivity of these lands in terms of forage grass, fodder leaves, and fuel wood has been reduced to almost naught. There are a number of bad consequences of mismanagement of community lands and faulty use of privately held lands. Increased soil erosion, rapid

*The figures in parentheses refer to the number of references cited at the end of the chapter.

siltation of water courses and reservoirs and recurrent floods are only a few examples of the more serious ones.

Although a number of important rivers and their tributories have their origins in this region, water for irrigation is not available in most parts of the region and often is not easily available even for drinking and other household purposes. High lifts, undulating topography and lack of appropriate water harvesting technology are some of the important reasons why very little use of surface water is made in the region. Ground water is out of reach because of hard rocky strata. As a consequence, only about 10 per cent of total net area sown in the region is irrigated and the rest about 90 per cent is rainfed.

The major foodgrain crops grown in the region are paddy and *madua* (ragi) in Kharif and wheat in rabi. Among the cash crops grown in the region chillis, potatoes, cauliflower, cabbage, capsicum, french beans and tomatoes are important. The use of purchased farm inputs such as high yielding seeds, chemical fertilizers and plant protection chemicals is very low and consequently, the yield rates are also low relative to those in the plains.

Animal husbandry is an important sub-sector of agriculture in the region. According to the Livestock census, 1972-74, there were about 2.11 million cattle, .64 million buffaloes, .74 million goats and .35 million sheep in the region (10). The cows and buffaloes are all of local non-descript type. The productivity of milch animals is very low—about 2 litres per day per buffalo and one litre per day per cow. The buffaloes are generally stall-fed and cows are mainly pasture-fed and partly stall-fed. There are no institutional arrangements for supply of feed, disease control and marketing of milk in the region.

There is wide-spread poverty, unemployment and under-employment prevalent in the region. This in conjunction with lack of basic amenities of life has resulted in large scale migration of people outside the region in search of means of livelihood. According to a RAPRAP survey conducted in Bhikiasen block, Almora district, about 50 per cent of the households had one or more family members working outside the region and 70 per cent of the migrants were in the age group of 25-39 years. Another important characteristic of hill agriculture is that the most of the farm operations, almost all excepting ploughing, are done by women. This has important implications for organizing training and extension education service in the region.

There is wide-spread prevalence of under-nourishment and diseases associated with it particularly among the infants and expectant and lactating mothers. Under-nourishment and diseases sap the vitality of the people to the extent that they are rendered incapable of doing any physical work. This is a serious wastage of human resources which if optimally conserved, developed and utilized could contribute significantly to the economic development of the region.

The region is poorly provided with both basic and supporting infrastructural facilities and services such as input supply system, institutional credit facilities, processing, storage and marketing facilities, extension education service, transport and communication and rural electrification. The absence or inadequacy of these services and facilities not only has hampered the modernization of agricultural but also has discouraged the entrepreneurs in establishing agro-based industries in the region. As a consequence, the potential of agro-industrial development present in the region remains, by and large, untapped.

The existing social and economic institutions and organizations in the region are also not geared to meet the needs and challenges of a growing economy. They are so structured as to keep people perpetually in the vicious circle of low output—low income—low consumption—no savings—no investment—low productivity—low output. There is need to over-haul these institutions and organizations so they could help speed up the process of economic development in the region.

3. Review of Past Strategies

The new strategy for ecological development outlined in this chapter is an outcome of long evolutionary process of planning and development in India. Therefore, a brief but critical review of the major approaches to rural development followed in the country in the past seems in order. A systematic beginning to promote rural development was made in October 1952 by launching the community Development Programme (CDP). Although CPD had made its impact felt by creating social over-heads in the rural areas including an extension machinery, it has not succeeded in generating community efforts to match those of the government for initiating and fostering a process of self-sustaining growth in the rural areas. Part of the reason for this failure was that the limited resources—financial, material and personnel—were spread too thinly all over the country side to make any tangible impact status of the people. This draw-back of CDP was sought to be removed by initiating package programmes like the Intensive Agriculture District Programme (IADP), Intensive Agriculture Area Programme (IAAP); and the High Yielding Varieties Programme (HYVP) in the decade of 1960's. Under these programmes resources were concentrated in areas which were responsive to such increased level of resource use due to their relatively better natural resource endowments and well developed institutions and organizations.Although these programmes helped usher in the 'Green Revolution' by contributing to rapid increase in agricultural production particularly foodgrains, they were criticized for having accentuated the intra-regional and inter-regional disparities in agricultural and economic growth. In view of this criticism, special programmes like the Small Farmer Development Agency (SFDA)/Marginal Farmer and Agricultural Labour (MFAL) development agency were initiated during the Fourth Plan period (1969-74) with main emphasis on the development of the weaker sections of the rural population so as to reduce the intra-regional disparities. Also, area development programmes like the Draught Prone Area Programme (DPAP), the Hill Area Development (HAD) and the Command Area Development (CAD) were taken up to remedy the inter-regional disparities.

Experience of all these programmes has shown that mere project approach or sectoral approach is not adequate to bring about over-all development of an area and to ensure that the benefits from development are equitably distributed among various sections of the society. Since the incidence of poverty and unemployment and the potential for economic development vary from region to region, an area approach based on considerations of local resource endowments and other peculiar features of the area concerned is more appropriate than a centralized approach imposed from above. A legitimate charge against India's planning process is that it has been pre-dominantly macro-oriented emphasizing national goals and priorities. ignoring the actual economic needs at the grass-root level and not drawing the requisite degree of peoples' participation in implementation.

The need for decentralization of planning was recognized in the Fourth Plan and a beginning was made in the direction of extending planning to the State, region and district levels (3). The Fifth Plan emphasized the importance of area/regional approach to development planning particularly in backward areas. The Sixth Plan strategy of rural development is decentralized micro-level block planning for full employment (1). Following this strategy, the Government of India launched the Integrated Rural Development (IRD) programme in 16 selected districts in the country in 1977-78. However, the methodology of preparing integrated area development plans is not yet fully developed. What is needed is a set of guiding principles for formulating and implementing integrated rural area development plans. In the following section, such a set of principles is presented.

4. Guiding Principles

In the light of the special characteristics of the region and the experience with the development programmes implemented there, the following principles seem appropriate as guidelines in selecting, formulating and implementing rural development programmes :

(*a*) Substantial improvement in the quality of life and purchasing power of the rural poor should be the sole criterion of selection of programmes and projects.

(*b*) Increased productivity through use of new technologies that are profitable, labour intensive dependable and ecologically sound should be the major means of increasing purchasing power of the people.

(*c*) A particular activity or enterprise in the area should be identified that could offer a starting point for the people with common interests and problems and that could break the vicious circle of low output—low income—no savings—no investment—low productivity—low output. All the necessary resources and efforts should be devoted to make a success of this activity (big push theory).

(*d*) After the vicious circle is broken and people have gained self-confidence, other promising programmes and projects which will catalyze self-propelling and self-replicating growth should be developed and launched with the help and cooperation of the local people concerned.

(*e*) The drain of brain and resources from rural areas should be minimized by providing organized farm and non-farm job opportunities within the areas themselves.

(*f*) The important role of women in the process of development should be recognized and nutrition, education, training and extension programmes should be so designed as to reach out to them.

(*g*) The rural producers particularly the rural poor should be organized on the lines of the 'Anand Pattern' dairy co-operatives so they could help themselves, make their presence felt, their voice heard and their requirements met.[1]

(*h*) The pattern of administration of programmes and projects should be such that it could provide opportunities for every development agency—official and non-official to make a contribution.

5. Specific Programmes and Projects

To tap the development potential of the region and to alleviate its major problems in accordance with the guiding principles of rural development outlined in the preceding sections, the following programmes and projects are proposed for the region :

1. The 'Anand Pattern' of dairy co-operatives has four basics namely, (*a*) a three-tier structure; (*b*) adoption of Anand bylaws; (*c*) continuous and concurrent audit; and (*d*) integration of production, processing and maketing activities.

5.1 *Land and Water Resources Development and Conservation Programme*

Land and water are two of the most important natural resources in the hill region having significant bearing on the level and pace of agricultural development. Both are subject to gradual deterioration over time as a result of mismanagement. Immediate attention is required to develop and conserve these resources so they could contribute more significantly to the objective of increased crop and livestock productivity. Proper management requires knowledge and application of appropriate soil and water management techniques. Apporpriate technology is now available with the Central soil and water conservation Research and Training Institution, Dehra-Dun and the G.B. Pant University, Pantnagar. The economic viability of some of these Soil and Water Conservation measures has already been established (4). What is needed is a coordinated action programme to educate, motivate and enable the farmers to adopt the recommended measures. Since the social benefits from soil and water conservation are greater than their private benefit, conservation subsidies should be provided to the adopter-farmers so they could capture part of the social benefits and thus sufficient economic incentive to adopt the necessary conservation measures.

It has been observed that availability of irrigation water itself leads to improvement in terraces and levelling of fields by the farmers with a view to use limited water facilities more efficiently. Therefore, wherever possible additional irrigation facilities should be provided in conjugation with soil conservation subsidies to that the complementary relationship between irrigation and improved terraces could be exploited to increase crop yields. In some areas, farmers are cultivating hill side having as steep slopes as 40 per cent or more. Needless to emphasize, cultivation on such steep slopes must be banned legally through mandatory land use regulations. However, in our enthusiasm to promote it, we should not forget that soil conservation in itself is not the objective but is only a means to increased crop and livestock productivity.

As a measure of land development, consolidation of scattered land holdings is also urgently needed to reduce wastage of labour and power in unnecessary movement from one location to another and to make adoption of improved soil and water conservation and crop production technologies more profitable to the adopter-farmers. A major argument against consolidation of land holdings in the region is that because of extreme variations in soil quality and fertility associated with altitude, slope and aspect, it would not be possible to do justice to each and every farm household in terms of keeping the size and quality of their land holding intact. This argument sounds to be true to some extent, but the gains to the farming community from consolidation would more than offset the sum of losses to individual members of the community, who could then be collectively compensated for their losses.

5.2 Integrated Livestock and Fodder Production Programme

Livestock occupy an important place in the economy of the region and there is tremendous scope for increasing livestock production particularly milk through improved breeding, feeding and disease control measures. Availability of fodder could be substantially increased by establishing new pastures on community lands, by planting dual purpose fodder-*cum*-fuel wood trees on community lands, and by rotational and controlled grazing of the existing pastures. Suitable species of pasture grasses and of fodder-*cum*-fuel wood trees are now available for use in the region. Economic viability of establishing improved pastures and of planting dual purpose trees on community land has also been established (4 and 9). What is needed is a co-ordinated action programme designed to increase fodder availability by utilizing the community lands and to increase milk production by improved feeding and effective disease control measures. As a long term measure, genetic improvement of the indigenous, non-descript and low producing cows and buffaloes through cross-breeding should also be taken up concurrently. The RAPRAP experience had been that natural services using pure Murrah bulls was a good method for improving the local buffalo stock and artificial insemination using the Jersey bulls' semen for improving the cows (6).

The RAPRAP experience also showed that in the absence of a regular and efficient system of supply of feed, provision of animal health care and marketing of milk, no milk production programme would survive and run smoothly and successfully. To take care of all these three prerequisites for a successful milk production programme, the milk producers could be organized on the 'Anand Pattern' of dairy co-operatives. The National Dairy Development Board provides the necessary technical guidance and assistance and the Indian Dairy Corporation, the necessary financial assistance for organizing milk production, processing and marketing on the 'Anand Pattern'. The author is convinced that organized milk production could well break the vicious circle and be the major catalyst in the process of rural development in the region.

Besides milk production, fattening of male buffalo calves for meat purposes could also be a very remunerative activity in the region. There is a lot of demand for buffalo meat in India and abroad particularly in the Middle-East countries. Here again, the problems of supply of feed, disease control and marketing could be tackled by the 'Anand Pattern' village co-operatives.

Another promising activity which could substantially add to the income and employment of the rural poor in the region is poultry. Its technical, financial and social feasibility has already been established under RAPRAP. Like milk production programme, a successful poultry production programme also would require institutional arrangements for supply of feed, provision of disease control measures and marketing of eggs and broilers.

Although, there is no concrete basis to support it, it is generally believed that the livestock population in the region is too much to be supported by the available fodder resources. Even if we ignore this controversial issue, in RAPRAP survey it was found out that only about 43 per cent of the total number of cows and buffaloes in the study area were in milk and the remaining about 57 per cent were dry (5). A sizeable proportion of the dry animals is completely unproductive and therefore, should be culled out. One way of motivating the farmers to reduce the number of un-productive animals is to regulate the number of animals that can be permitted to graze in the community pastures by levying a grazing tax per head of animal. Besides the tax, an upper limit on the total number of animals that a household can be permitted to send for grazing should also be set by mutual concensus of the members of the community. Besides reducing the number of uneconomic animals, this type of regulation would also generate additional revenue to the village community which could be used for providing basic village amenities like drinking water, foot paths, panchayat ghar, etc.

5.3 *Vegetables Production Programme*

The region enjoys higher comparative economic advantage in producing off-season vegetables in summer and rainy seasons than the plains. Since there are virtually no supplies of fresh vegetables like tomatoes, cauliflower, cabbage, capsicum, french beans, peas, etc. from the plains in summer and rainy season there is a lot of demand for these vegetables and the prices fetched are high indeed and offer sufficient incentive to the producers to increase their output. But limited availability of irrigation water is the most serious constraint on increasing acreage under vegetables and the possibilities of this constraint being relaxed in the immediate future are highly remote. Alternatively, production of vegetables could be increased to some extent by introducing improved varieties alongwith other complementary inputs and practices.

With out-put of vegetables remaining the same, income from vegetables could still be increased substantially by improving the existing marketing system.

Vegetable growers' co-operatives, if organized and operated on the 'Anand Pattern', could play an important role in improving both production and marketing of vegetables in the region.

Besides production of fresh vegetables, production of vegetable seeds could also be an important source of income to the farmers in the region. The climate of the region is idealy suited to production of good quality seeds of several European types of vegetables. A separate project for this purpose could be prepared and launched in suitable areas in the region.

5.4 *Fruits and Nuts Production Programme*

The agro-climatic conditions obtaining in the region permit profitable production of a wide variety of temperate and sub-tropical fruits. Among the temperate fruits, apples, apricots, peaches and plums, and among the nuts, walunts, peaunts, chestnuts and hazelnuts could be successfully grown in the region. Sub-tropical fruits like lemons, limes, loquates, mangoes and litchis can also be grown at suitable locations. Improved technology for production of most of these temperate and sub-tropical fruits is now available with the Horti-culture Research Station, Chaubatia, Almora District. However, for a number of reasons including non-avail-ability of suitable land in large compact blocks, lack of long term credit facilities needed for plantation of fruit orchards and lack of knowledge about scientific management of orchards, fruits production sub-sector still remains insulated from the impact of new technology. Thus, there is need for modernization of fruits production sub-sector by introducing new techno-logy through a well thought-out and phased programme.

The fruits production programme would consist of (*i*) rejuvenation of old orchards through scientific management, and (*ii*) establishment of new orchards using the latest available technology. For establishing new orchards in the privately held lands, consolidation of holdings would be necessary. However, it is mainly the community lands where new orchards would be feasible to establish on any meaningful scale. Both production and marketing of fruits would be improved substantially by organizing fruit growers' co-operatives on the 'Anand Pattarn'.

5.5 Food grain Crops Production Programme

Since the primary purpose of almost all farmers in the region is to produce mainly for meeting family consumption requirements, the main thrust of this programme should be to increase the per hectare yield rates of staple food crops like paddy, *madua* and wheat by introducing high yielding varieties along with other complementary inputs and practices. It has been demonstrated by the scientists of G.B. Pant University working under RAPRAP that is possible to substantially increase yield rates of all major foodgrain crops in the region by the use of new technology now available with the University and the Vivekanand Laboratory for Hill Agriculture, Almora. Economic viability of new techonology has already been tested under RAPRAP by conducting trials and demonstrations in the farmers' fields. To examine the economic consequences of adoption of new crop and livestock production technologies, a set of alternative optimal farm plans was prepared by the linear programming technique for a typical farm situation in the Naurar watershed in

Almora district (7). The optimal farm plans are presented in Table 7.1 and their economic consequences in Table 7.2. It is evident from Table 7.2 that when resources were merely recombined according to profitability criterion under A_1, returns over variable cost (ROVC) increased by about 32 per cent over the existing situation. The highest ROVC was attained under A_6 when improved crop production technology was introduced along with milch animals, poultry and the credit constraint was relaxed. A similar exercise done for a small sub-watershed showed that ROVC could be increased by about 89 per cent if new crop technology is used in place of the traditional one (8). All these exercises showed substantial potential for increasing farm production, income and employment in the region. It remains for the action agencies to tap this potential by formulating and implementing appropriate programmes and projects.

5.6 *Social Forestry Programme*

As mentioned in section 3, about one-third of the total geographical area in the region is held under one or the other form of community ownership. These lands are at present almost barren and badly eroded. However, with proper management they could be used for putting up pastures grasses and fodder and fuel wood trees. A study conducted in the Ramaganga Catchment has shown that it is financially feasible to plant chir, bhimal, and chir+bhimal trees on these community lands (9). The net benefit annuities, gross benefit-cost ratios and financial internal rates of returns for these three types of plantations are presented in Table 7.3. Besides meeting fodder and fuel wood requirements of village communities, the social forestry programme would also generate substantial employment opportunities within the region. According to a rough estimate, the programme could generate additional employment at the rate of 60-80 man days per hectare per annum. Therefore, there is urgent need for formulating and launching a social forestry programme in the region with the co-operation and help of the village communities concerned.

5.7 *Agro and Forest-Based Industries Programme*

Because of relatively small land base, agriculture even at its best will not be able to provide gainful employment to all the job-seekers in the region. Therefore, alternative avenues of providing productive employment will have to be found out outside the agricultural sector. One such avenue having substantial employment potential is the agro and forest-based industries suited to the region. Some of the agro-based industries which could be taken up successfully in the region are milk processing units, fruits and vegetables processing units, packing case making units, furniture making units, rope and basket making units, straw board making units, pharmaceutical units, etc. This programme would operate successfully only if these industries are organised in the cooperative sector restructured on the 'Anand Pattern'. The producers' co-operatives could play a very important role by helping their members in procuring raw materials at reasonable prices, in getting needed financial assistance from institutional sources and in marketing their produce. The programme should also provide for training of rural artisans.

5.8 *Human Resource Development Programme*

Since labour is the most important source of income to most of the households in the region, special

TABLE 7.1

Alternative Optimum Plans for a Synthetic Farm in Naurar Watershed (Size of Farm = 0.86 ha)

Crop/Livestock/Poultry	*Existing Crop Pattern*	*Alternative Plans* (per cent of total cropped area)*					
		A_1	A_2	A_3	A_4	A_5	A_6
1. Local paddy + jhungra	19.41	22.17	11.65	11.65	11.65	Crop pattern under A_5 and A_6 is the same as under A_3.	
2. High yielding paddy	0	0	11.94	8.37	0		
3. Ragi + bhat/urd	19.83	22.12	18.64	18.64	35.50		
4. Capsicum	0	0	0.21	0.58	0		
5. Chillis	1.29	3.49	1.16	1.16	1.16		
6. Tomatoes	0	0	0.58	0.58	0.50		
7. French beans	0	0	1.16	1.16	1.16		
8. Soyabeans	0	0	4.66	4.66	0		
9. Summer potatoes	0	0	0	2.33	0		
10. Rabi potatoes	0.15	5.82	0	5.24	0		
11. Local wheat + barely/lahi	21.55	22.16	0	0	0		
12. High-yielding wheat	0	0	19.50	22.92	7.20		
13. Peas	0	0	3.49	0.58	3.49		
14. Oats	0.14	0	0	1.16	0		
15. Maize	2.11	0	0	0	0		

16. Rabi fallow	35.52	24.24	18.64	20.97	18.64		
17. Land left unused (slack activity	0	0	8.37	0	20.70		
18. Local buffalo (No.)	1	0	0	0	1	1	1
19. Improved cow (No.)	0	0	0	0	1	1	1
20. Poultry units (1 unit = 50 layers)	0	0	0	0	0	0	2

*A_1 = optimum crop plan with existing technology; A_2 + optimum crop plan with improved technology and limited credit; A_3 = optimum crop plan with improved technology and unlimited credit; A_4 = optimum crop-*cum*-livestock plan with improved technology and limited credit; A_5 = optimum crop-*cum*-livestock plan with improved technology and unlimited credit; and A_6 = optimum crop-*cum*-livestock-*cum*-poultry plan with improved technology and unlimited credit.

TABLE 7.2

Returns Over Variable Cost, Marginal Value Products of Scarce Resources, Employment Potential and Credit Requirements under Various Alternative Farm Plans in Naurar Watershed

	Existing Situation (1973-74)	*Alternative farm plans*					
		A_1	A_2	A_3	A_4	A_5	A_6
1. Returns over variable costs (ROVC) (Rs.)	1,813	2,385	3,634	4,124	3,972	5,353	3,103
2. Percentage increase in ROVC over the existing situation	—	31.54	100.40	127.47	118.53	195.25	346.93
3. Marginal value products of scarce resources							
Kharif irrigated land (Rs./ha)	—	1,415	3,804	3,890	1,779	3,890	3,890
Kharif rainfed land (Rs./ha)	—	2,045	1,996	1,925	1,779	1,925	1,925
Rabi irrigated land (Rs./ha)	—	1,351	2,966	8,199	2,966	8,199	8,199

Rabi rainfed land (Rs./ha)	--	1,544	0	955	0	955	955
Human labour during October-November (Rs./day)	0	0	0	5.60	0	5.60	5.60
Kharif cash (Rs.)	0	0	1.56	0	4.76	0	0
Rabi cash (Rs.)	0	0	2.70	0	2.70	0	0
4. Employment potential (mandays used per annum)	230 (19.18)*	308 (25.74)	515 (43.04)	562 (46.96)	312 (26.09)	632 (52.81)	665 (55.57)
5. Credit requirement (Rs./ha)	0	0	233	868	233	1,287	4,830

*Figures in parentheses indicate percentage of total man-days available per annum.

TABLE 7.3

Net Benefit Annuities, Gross Benefit Cost Ratios and Financial Internal Rates of Returns for Chir, Bhimal and Chir + Bhimal Plantations in Ramaganga Catchment

Type of Plantation	Net benefit annuity (Rs./ha)			Gross benefit cost ratio			Financial Internal rate of return
	Discount rates (%)			Discount rates (%)			
	5	10	15	5	10	15	
1. Chir	1,019	105	0	3.54	1.29	0.51	12.16
2. Bhimal	591	314	77	2.21	1.64	1.15	17.27
3. Chir + Bhimal	978	130	0	3.38	1.34	—	12.58

programmes for proper development, conservation and utilization of this resource should be formulated and implemented. For proper physical and mental development, proper nourishment and appropriate education are needed. An effective and efficient public health system and a public nutrition programme could help conserve the human resources. The hills people have already realized the importance of education as a means of increasing their family incomes. This is reflected in the relatively high literacy rate (25.5 per cent) in the region. However, the appropriateness of the education is as much questionable here as anywhere else in the country. It only prepares the children to leave the rural areas and get jobs in urban areas.

Programmes for improving health and nutritional status could be integrated with programmes for increasing production of milk, fruits and vegetables and food-grains. There should be a special nutrition programme for the vulnerable section of the society, *i.e.*, the children below 14 years of age and expectant and lactating mothers.

For proper utilization of that part of labour force which cannot be absorbed locally in the agricultural and allied activities, there should be a government sponsored labour employment assistance service in the region. This service should provide information about availability of jobs within and outside the region, assist the job-seekers in getting jobs, and arrange for pre-job training, if necessary. The service should keep in touch with all prospective employers including public, private and co-operative establishments in and outside the region. At present labour contractors exploit the poor, ignorant and unorganized job-seekers. Labour co-operatives could be organized to minimize this exploitation and to improve the bargaining power of the labour.

5.9 Supporting Infra-structural Services and Facilities

Successful implementation of the various production-oriented programmes mentioned in the preceding paragraphs will need creation of new and streamlining and strengthening of the existing infra-structural services and facilities. First of all, a specially designed extension education programme will need to be developed and launched in the region to transfer whatever new appropriate technologies are available with the research institutes doing research on problems of hill agriculture. At present both the State Government and the G.B. Pant University, Pantnagar are involved in extending extension services to the farmers in the region. Although a mechanism exists to ensure that there are no discrepancies and contradictions in the new scientific knowledge that the State Government and the University extension workers transfer to the farmers, instances of such discrepancies and contradictions do exist. Furthermore, none of the two agencies, has taken into consideration the place and role of women in hill agriculture in planning their extension education programmes. Special attention will need to be paid to educate the hill women in improved crop and livestock production techniques, in improved household management, nutrition and childcare. This will necessitate employment of women extension workers.

Another important feature of the reorganized extension service should be adoption of 'whole farm approach' in place of commodity or mono-disciplinary approach. Short term courses to train young farmers including women should be organized by the University at regular intervals. Besides, short-term in-service training courses for extension workers once each before the start of kharif and rabi seasons should also be organized by the University. It would be desirable to have a unified extension education service under the control of the University.

Adoption of new technologies by farmers will require injection of substantial amounts of outside capital into the economy of the region. According to a RAPRAP study, a typical farm of 0.86 hectare would need credit assistance of the order of Rs. 4,830 per hectare in order to be able to adopt the optimal crop-*cum*-livestock-*cum*-poultry plan (7). It is doubtful whether any nationalised commercial bank would grant that much credit to a hill farmer, even if the projected repayment capacity of the farmer justifies it. This means that some other alternative source of credit will need to be found out. Village credit co-operatives organized on the 'Anand Pattern' could be a possible alternative and deserve trial on a small scale.

Institutional arrangements for supply of various farm inputs at right time, in adequate quantities and at reasonable price is another prerequisite to adoption of new farm technologies. The RAPRAP experience shows that in the absence of such arrangements, farmers are not able to adopt the new technologies on a sustained basis. Here again, the answer is the 'Anand Pattern' of village co-operatives.

The RAPRAP experience also shows that the existing system of marketing of fruits and vegetables is far too inadequate and inefficient to provide any incentives to the producers to expand their output. The producer's share in the consumer's rupee is very low and a lion's share is taken away by the intermediaries. To enable farmers to get a fair share in consumer's price and to motivate them to produce more, it would be necessary to organize fruit and vegetable producers on the 'Anand Pattern'.

Lastly, the provision of such other basic infrastructural facilities and services as rural roads, storage facilities, market yards, market intelligence and rural electrification would also help increase farm production and income substantially. Of course, the provision of these facilities and services is made as part of the 'Minimum Needs' programme. There is need for expediting the provision of these facilities in areas where their absence is really a major obstacle to output expansion.

6. Planning and Implementation Machinery

So far we have briefly discussed what should be done to initiate and foster the process of rural development in the region. One crucial question that still remains to be answered is who will be responsible for planning and implementation of these development programmes. One of the major reasons for failure of most rural development programmes in India has been the multiplicity of government departments and agencies engaged in rural development and lack of co-ordination among their activities. Besides, traditional jealousies among various government departments also stand in the way of mutually beneficial interaction and communication among them. It is proposed that a separate autonomous body which may be called the Uttar Pradesh Hill Development Authority may be established for the purpose of planning and implementation of various development programmes. The services of all the State Government officials engaged in the task of agricultural and rural development in the region may be put at the disposal of the Authority. The authority should have full financial autonomy and should be charged with the responsibility of (*a*) development policy formulation; (*b*) planning; (*c*) devising and prescribing implementation measures; (*d*) programme implementation; (*e*) budgeting and financial management; (*f*) monitoring and concurrent evaluation; and (*g*) integration and co-ordination. The head of the authority must be a professional development planner having requisite experience in managing inter-disciplinary rural development programmes and must enjoy all the powers of all the heads of departments concerned with rural development in the region.

VII Summing Up

The hill region occupies an important place in the economy of the State of Uttar Pradesh. In the absence of an appropriate economic development policy, the region still continues to be economically depressed and has been by-passed by the recent wave of technological break-through in agriculture. An appropriate development strategy is needed urgently to promote optimal development, conservation and utilization of natural and human resources of the region and thereby to speed up the pace of agricultural and rural development. An integrated area approach seems to be the most appropriate strategy for economic development of the region. With the adoption of new technologies that are now available in the fields of soil and water conservation and crop and livestock production, it is possible to increase farm output, income and employment substantially in the region. To facilitate wide-spread use of new technologies, the existing infrastructural services and facilities such as input supply system, credit system, marketing system, extension education system, etc. will need to be streamlined and re-organized. The rural poor in the region need to be organized on the 'Anand Pattern' of producers' co-operatives to help themselves, to protect their interests and have their legitimate needs fulfilled. At present there is no single agency in the region that has both the authority and the scope to perform the essential functions of programme planning, programme implementation and integration and co-ordination of development programmes. It is proposed that an autonomous authority, which may be called the Uttar Pradesh Hill Development Authority, may be created to perform these functions.

REFERENCES

1. Azad, R.N., IRD : Concepts, Objectives and Strategies, Indian Farming (Integrated Rural Development Number), Oct.-Nov. 1978, p. 39.
2. Hardin, Garrett, "The Tragedy of the Commons" *Science*, Vol. 162 (13 Dec. 1968), pp. 1243-1248.
3. Planning Commission, Fourth Five-Year Plan, 1969-74, New Delhi, 1970, p. 15.
4. Prasad, Shambhu, Economics of Some Selected Soil Conservation Measures in Ramganga Catchment in U.P. Hills, Unpublished M.Sc. (Ag.) Thesis, Deptt. of Agri. Econo., G.B. Pant University of Agriculture and Technology, Pantnagar, 1979.

5. Shah, S.L. *et al.*, Fragmentation and Scatteredness of Holdings, in Rural Area Development (Book), G.B. Pant University of Agriculture and Technology, Pantnagar, 1976, p. 59.
6. Sharma, O.P., *et al.*, Animal Science, in Rural Area Development (book), *ibid.*, pp. 209-10.
7. Singh, Katar, Some Alternative Strategies for Optimal Use of Land and Others Resources in the Hill Region of Uttar Pradesh, Indian Journal of Agricultural Economics, July-Sept. 1977, Vol. XXXII, No. 3, pp. 113-120.
8. Singh, Kartar and K.M.B. Rahim, Identification and Evaluation of Optimal Cropping Systems for a Typical Watershed in Uttar Pradesh Hills, Indian Journal of Agricultural Economics, Oct.-Dec. 1978, Vol. XXXII, No. 4, pp. 29-35.
9. Tewari, D.D., Financial Feasibility of Afforestation of Community Lands in Ramganga Catchment, Unpublished M.Sc. (Ag.) Thesis, Deptt. of Agri. Econ., G.B. Pant University of Agriculture and Technology, Pantnagar, 1980.
10. Uttar Pradesh Forest Department, A Project on Afforestation, Integrated Watershed Management, Torrent Control and Land use Development in the U.P. Himalayas and Siwaliks, Lucknow, 1978, p. 21.

8

Bio-Geographical Resources Their Environmental Degradation and Conservation—A Case Study of Garhwal Himalaya

P.S. NEGI AND DEVENDRA PAL

I. Introduction

The Garhwal Himalaya encompasses five out of eight districts of U.P. hills, constitutes 81.7 per cent hilly area, showing representation of complex phenomenon like vegetation in the form of forest, grasslands, marshers, alpine pastures and meadows with respect to change in topography, geology, soil and other natural resources with climatic factors. The scenic mountain belt area abounds in fertile valleys, deep gorges, rugged ravines and covering all the three section of the Himalaya, *viz.*, outer sub-Himalayan zone in form of Siwalik foothills, middle Lesser Himalaya and high mountainous horned sharp edged peaks in the Greater Himalaya. The

Garhwal Himalaya has a large number of glaciers which provide water to a large number of major streams and their tributaries. These are Alakhnanda, Bhagirthi, Nandakini, Dhauliganga, Mandakini, Yamuna, etc. There are a large number of water falls ranging from a few metres to hundreds of metres. Nanda Devi, Chaukhamba, Trisul, Banderpunch, Badrinath, Kedarnath, Hemkund, Rupkund are located within the Garhwal region which are important pilgrimage shrines of Hindu and Sikh religions. Nanda Devi National Parks, Corbett National Park, Rajaji National Park, and Musk Deer Sanctuary etc. are located within this region. It lies between the latitude 29°26′—30°28′N, and longitude 77°49′—80°6′E, falling between elevation of ca 100 m. —7000 m. and covers an area 30,090 sq. km. supporting a population of 24,29,631 (1981). According to the altitude, vegetation, soil, topograhy, direction and extent of sunshine, there are particular types of land use, human settlements and culture in the villages and small townships sustaining lower population as compared to area (Lal *et al.*, 1987).

The river Tons marking its Western limit separates it from Himachal Pradesh and district boundaries of Nainital, Almora and Pithoragarh separate it from the Kumaun in the East and its limit in North coincides with the international boundary with Tibet.

In the present work the authors have dealt with flora *i.e.*, the forests, medicinal plants and herbs. The wild life resource has not been included in this work.

2. Climate

The different types of climate from sub-tropical to cool temperate types are observed in the area mainly because of marked variation in altitude from the plains in the South to snow clad ranges in the North. Winter are very severe and snow fall is caused by winter depression commonly during December, January and February which normally comes down upto 1500 m. however, in 1945 and 1980, it came down exceptionally low in Doon Valley upto Rajpur 975 m. The minimum and maximum temperature ranges between 5°C and 46°C in the region. The annual rainfall ranges from 700 to over 2000 mm., the rainy season is flanked with a warm and cool dry period. The area is markedly affected by summer monsoon arrival which heralds the onset of rainy season in later part of June, and its retreat by the end of September coincides with lowering of temperature and gradual advent of the winter season. The month of October is quite pleasant all through out the area. Winter showers are regular feature during December-February.

3. Soil

Soils of the area vary in their texture, morphology, moisture content and chemical composition from slope to slope and valley to valley depending upon ecological setting, geological formations, provenance of material etc. In tropical zone (900 m.) the alluvial soils mixed with boulders, gravels and pebbles are commonly found within the valleys in the Siwalik Hills, Duns and Dhabar regions. The brown forest soils generally varying from loam to clayey loam are fairly widespread in the warm temperature and the sub-tropical zones. In the cool temperate and cold region (1800-3000 m.), brown deciduous forest soils, which owe their origin to glacial and fluvio-glacial processes. The alpine zone extending above 3000 m. has soil of glacial origin with barren rocks due to frost action and meadows. In general the soils of the area are quite thin, stoney and poor in fertility except in case of river terraces, valleys, Duns, Bhabar and Tarai areas where these soils are transported and deposited in the present locations

4. Vegetation and its Geographical Distribution

The forest dominate the phytogeography and serve as a most valuable natural wealth and ecological stock not only for the area but to entire country. The flora is correspondingly different types in different parts as it is characterised by change in climate and other topographical features etc. The national forest policy revised in 1952 laid down the emphasis that there should be 60 per cent forest cover in the Himalayan and other mountainous terrain and 20 per cent in plains. Though U.P. forest statistics (1981, p. 13, Lucknow) show that 76.87 per cent of study area is under forest but out of this only 41 per cent is estimated to be really covered with forest (Joshi, D.P., 1979). The satellite imageries of U.P. Hills indicate 47.5 per cent forest area (Bahuguna, S.L., 1978) while Desh Ka Paryavaran (Environment of country—a book published by Gandhi Peace Foundation, New Delhi, 1983) reveals that this area has shrunk to 37.45 per cent and even half of this area contains very less forest density. The U.P. State is having 17.24 per cent area under forest (S. Paul, energy from forest) while India comprises 22.8 per cent area under forest (India's forest 1980, Central Forestry Commission). Puri (1960), Champion and Seth (1968) have studied the complicated structure of vegetation with respect to different topographical aspects in details. The area may be divided into three broad vegetational zone, mainly on the basis of climate, altitude, with some over-lapping of transitional factors. However, there are practically no sharp boundaries between these vegetational zones due to a combination of factors like topography, soil and geology.

4.1 Lower or Tropical and Sub-Tropical Zone

This zone covers Siwalik ranges and valleys of lower zone or sub-Himalayan tract and is the extension of sub-deciduous forest belt of Bhabar areas. It ranges roughly between 300-1200 m. in height and is dominated by deciduous and sub-deciduous types. *Shores robusta* (Sal) (Plate 1a) is the most common species found generally upto the elevation of about 750 m. on Southern slopes and 1200 m. on Northern slopes. *Dalbergia sisso* (Shishum) are dominant along the river beds, while in more moist soils, dominating species are *Cedrela toona* (Tun), *Ficus glomerata* (Gular), *Eugenia jambolana* (Jamun). In isolated patches of grasses, there are trees of *Acacia catechu* (Khair), *Butea monosperma* (Dhak), *Bombax Ceiba* (Semal), *Adina cordifolia* (Haldu) and *Holoptelea integrifolia* (Kali papri), etc. In dry belt towards West, Sal is replaced by thorny succulent Euphoribias on slopes. The other important trees of this zone are *Aegle marmelos* (Bel), *Anogeissus Latifolia* (Bankli), *Mallotus philppensis* (Rohni), *Dendrocalamus strictus* (Bans), *Dillenia indica* (Chalta), *Engelhardtia spicata* (Mauwa), *Terminalia* spp. and *Tectona grandis* (Teak) etc. Some fresh-water swamp forest are constituted by *Alstonia scholaris* (Satni), *Bischofia javanica* (Kot Semla), *Carallina teractirata* *Litsaea chinensis* (chandna), *Syzygium cumini* etc. The ground vegetation like feorns and allies are poorly represented. Linas and climber include *Bauhinia vahlii* (Malu), vitis sp., and *Philodendron* sp; etc. *Pinus roxburghii* (chir pine) begins to appear at 1000 m. to 1600 metres.

4.2 *Temperate Zone*

This middle zone is dominated by coniferouse forests and approximately covers the areas from altitude of 1,200 to 2,000 m. with maximum precipitation both in summer and winter. *Pinus roxburghii* generally is replaced by *Pinus wallichiana* (Kail) from 1,500-3,100 metres. *Cedrus deodara* (Deodar) is quite abundant forming pure forest stand. At these altitudes *Quercus leucotrichophora* (Banj) with its common associate *Rhododendron arboreum* (Burans) also grows in separate patches. In inner Himalaya *Betula utilis* (Bhojpatra), *Salix* spp., and *populus* (popular) are abundant on certain soil types. At higher altitudes *Aeuculus indica* (Panger), *Quercus semicarpifolia* (Kharsu), *Q. Himalaya* (moru) along with conifer like *Abies pindrow* (Silver fir or Ronsli), *Picea morinda* (Himalayan Spruce or Roi), *Cupressus Torulosa* (Himalayan cypress or Devidiar), *Taxus baccata* (Thunner), are most common components of vegetation. *Rhododendron companulatum* (Burans-light blue or white flowers) grows at higher altitudes. In inner valleys of dry mountains, *Pinus gerardiana* (chilghoza pine) is also found, but it is not always the case every where and the species is some what local in the inner arid valleys of the Himalaya around Niti pass in Chamoli district. The other common tree species are *Acer caesium* (Kainju), *A. pictum* (Kainchli), *Buxus wallichianum* (Papri), *Corylus* spp. (Kapasi etc.), *Juglans regia* (Akhrot), *Rhus punjabensis* (Titri), *Ulmus wallichiana* (The Himalayan Elm or Emroi), mixed with *Pyrus pashia* (Mehal), *Myrica nagi* (Kaphal), *Lyonia ovalifolia*, *Fraxinus* spp., *Viburnum stillulatum* (Richhoi), etc.

The pine forest have poor under growth while other have a thick canopy of under shrub and herbs. Few important and commonly occurring shrubs are *Abelia triflora* (Mali), *Berberis* spp. (Kingore, Toter, etc.), *Coriaria nepalensis* (Mansuri), *spiraea* spp., (Latkar, Takoi), *Indigofera* spp., (Sakina, etc.) *Rubus* spp. (Hinser, etc.) *Myrsine africana* (Rikhdalmi), *Sorbus* spp., *Strobilanthes wallichii* (Janu), and *Rosa macrophylla* (Kuja), etc.

4.3 Alpine Zone

This zone representing unique ecological conditions, includes sub-alpine and alpine areas and ranges from 2,000 to 4,000 m. It marks the limit of tree growth at about 3,900 m. known as timber line, where the plants height gets considerably reduced. Plants are mostly dwarfed and shrubs and grasses are cushion shaped. At about 4,800 metres and above snow line, plants growth almost nil. On lower levels of this zone some *Rhodendron* sp., *Betula* sp., *Juneperus squamata* (Guggal), *J. macoropoda* (Dhup), *Lonicera purpurascens*, *Caragana brevispina* (Kathur), *Salix elegans* (Kadvi), *Cotoneaster acuminata*, (Raunshi), etc. and other are distributed in varying conditions. The alpine meadows are well known for their beautiful and attractive harbaceous plants species, occupy their maximum area in Chamoli and Uttarkashi districts. These meadows remain covered with snow during winter (November to April) and develop their flora from May to October. Most of the plant species possess medicinal value, *e.g.*, 250 medicinal herbs are found only at Tungnath (3,680 m.) peak in Chamoli district. These herbs generally have short period of vegetative growth and swollen roots due to food shortage while aerial part dies due to heavy snow. These include mainly species of *Primula*, *Patintilla*, *Polygonum*, *Geranium*, *Aster Anemone*, etc. The alpine herbs mostly belong to the families Ranunculacease, Papaveracease, Fumariacease,

Caryophyllaceae, Centalaceae, Primulaceae, Campanulaceae, Asteraceae, Scrophulariaceae, Lamiaceae, Orchideceae and others (Rau 1975, Semwal and Gaur 1981).

5. Resource Potential

Forests of the area serve basic requirement of fuel, fodder, timber and other forest products and represent a vast resource potential. District-wise distribution of forest area and their percentage with respect to geographical areas has been shown in Table 8.1.

The forest area and production of a few common plant species being exploited at commercial level is given in Table 8.2. This Table showing that Chir (Pine) have been exploited at largest extent for timber wood.

The rate of forest exploitation is increasing due to increased consumption of forest products. In U.P. hills each person utilises nearly 2 ton (estimated by international authority) fuel wood annually in addition to wood used in agricultural implements, furniture and timber etc. The fire wood consumed in the hill region of Kumaun and Garhwal alone would be 9,630,652 tons each year and to meet this huge demand, pressure on forest is increasing steadily.

A survey conducted by Forestry Department of Garhwal University in 1983 concluded that an average family requires 13-14 trees annually (2 as fuel wood, 1 as timber wood, 5 as fodder tree, 1 for agricultural implements and pots, 1 for fruit, flowers and shade, 1 for marriages ceremony and death, 3 for other purposes). A village of 50 families may utilize roughly 600-700 trees and condition of village forest of thousands year old

TABLE 8.1

District-wise Distribution of Forest Area

(*In Sq. Kms.*)

Sl. No.	*District Name*	*Under Forest Department*	*Civil-Soyam Forest*	*Panchyati Forest*	*Personal and municipal Forest*	*Total Forest*	*Percentage in Relation to Area*	*Per Capital Forest Area*
1.	Pauri	2,469.22	1,806.00	376.04	6.07	4,657.33	85.61	7.56
2.	Chamoli	3,704.86	1,043.00	521.49	—	5,269.35	57.74	16.07
3.	Uttarkashi	2,694.26	870.00	—	—	3,564.76	88.61	41.54
4.	Tehri	1,495.74	516.00	—	204.90	2,216.64	80.63	8.10
5.	Dehra Dun	1,954.89	148.00	—	—	2,102.89	71.78	3.03
	Total	12,318.97	4,483.00	—	—	17,810.97	76.87% Average	15.26 Average

Source : Forest Statistics, U.P. 1981, pp. 13 and 22-24 Lucknow.

TABLE 8.2

Forest Area and Production of a Few Common Species

Sl. No.	Plant species	Area in Sq. Kms.	Production (in Cubic metres round)		
			1976-77	1977-78	1978-79
1.	Chir	4,117.82	3,18,840	3,42,141	4,13,654
2.	Deodar	184.11	16,933	10,378	9,657
3.	Fir and Spruce	894.05	43,916	76,766	92,869
4.	Kail	184.17	10,714	6,019	6,849
5.	Cypress	32.95	—	5	2
6.	Oak	—	1,368	6,238	369

Source : Forest Statistics, U.P. 1981, pp. 27 and 51, Lucknow.

easily can be estimated. As the area is economically undeveloped and rural, inhabitants are depend on the forest in their vicinity for fodder, fuel and timber requirements. The average production of fuel wood (4,218^{3} m) is more than the timber (0.03^{3} m) while considerable timber wood is transported to outside areas.

The production of fuelwood is maximum in Chamoli district (1.6251^{3} m) probably due to cold weather and Uttarkashi produce maximum (0.6328^{3} m) timber wood being a prosperous in forest wealth and maximum per capita forest area *viz.*, 41.54 Sq. Km. (U.P. Forest Statistics 1981). The forest resources of the area have been ruthlessly exploited at larger scale for maximum revenue, social and economic development after independence. About 58.62 per cent total forest revenue of U.P. State is received from U.P. hills while it only covers 17.89 per cent geographical area of the State.

The pine forest which form main tree crop of the area is not only being exploited largely for timber wood but also for resin extraction. Resins which are important plant exudates used in adhesive, paints, medicine, cosmetics, textile and turpentine industries. Annual resin production in U.P. during 1976-77 was 2,10,156 Quintals, it was 1,41,241 Quintals in 1981-82.

The area is highest producer of resin in India and main quantity of it is sent out side to cater the needs of different industries. At present, 17 per cent of total U.P. hill annual income is obtained through forest while only 1-2 per cent of total annual income of U.P. is received from forest of U.P. The forest resources have been given very little attention from small scale industrialisation point of view. Only a survey of industries based on plants made by Raturi (1947) and later by Gupta (1960) for Tehri Garhwal was done. Authors have studied the forest flora and cultivated plants of the area for last four years and made possible efforts to explore it for small scale industries (Table 8.3).

TABLE 8.3

Economic Plants of Garhwal Himalaya for Small Scale Industries

1. Fibres for Ropes, Cordage and Stuffing etc.

Grewia oppositifolia (Bhimal), *Agave sissalana* (Banskera), *Helictres isora* (Marorphal), *Hibiscus sadariffa* (Patson), *Boehmeria rugulosa* (Genthi), *Cannabis sativa* (Bhang), *Linum usitatissium* (Alsi), *Musa* spp. (Kela), *calotropis procera*, *Saccharum munja* (Munj), *Andropogon* spp. (the spear grass), etc.

2. Gum Resin, Oil, Etc.

Pinus roxburghii (Chir), *P. wallichiana* (Kail), *Cedrus deodara* (Deodar), *Bauhinia retusa* (Semla), *Moringa pterygosperma* (Sanjan), *Butea frondosa* (Dhak), *Acacia* spp. (Babool, etc.). *Shorea robusta* (Sal), *Prunus* spp. (Aru, Khoomani, etc.), *Melia azedarach* (Bakain), *Anogeissus latifolia* (Bakli), etc.

3. Sport Goods

Morus alba (sahtoot), *Salix elegans* (Kadvi), *Pyrus pashia* (Mehal), *Adina cordifolia* (Haldu), *Flcourtia ramontchi* (Kango), *Juglans regia* (Akhrot), *Delbergia sissoo* (Shishum), *Tectona grandis* (Teak), *Butea monosperma* (Dhak), *Ficus racemosa* (Gooler), *Cedrela toona* (Tun), etc.

4. Wood for Carving and Engraving

Juglans regia, *Olea glandulifera* (Gald), *Alstonia scholaris* (Satain), *Wrightia tomentosa* (Dudhi), *Emblica officinalis* (Amla), *Buxus wallichianum* (Papari), *Aesculus indica* (Panger) *Acer oblongum* (Kirmola), etc.

5. Mats and Baskets

Arundinaria spp. (Ringal, etc.), *Salix elegans*, *Morus abla*, *Typha* spp. (Patera, etc.), *Grewia oppositifolia*, *Musa paradisica* (Kela), etc.

6. Soft Wood for Musical Instruments and Toys

Leycesteria formosa (Bhuj-nali) *Buxus sempervirens* (Papri).

7. Mine Frame Work

Quercus leucotrichophora (Banj), *Pyrus pashia*, *shorea robusta*, *Terminalia tomentosa* (Sain), *Acasia nilotica* (Babool).

8. Pen Holders and Pencil Slates

Cedrus deodara, *juglans regia*, *Cedrela toona*, *Pinus wallichianum* (Kail), *Bombax ceiba* (Semal), *Adina cordifolia*.

9. Sericulture

Morus alba, *Quercus leucotrichophora*, *Q. himalayana* (Moru), etc.

10. Bobbins and Shuttles

Adina cordifolia, *Carpinus Viminea* (Chamkharik).

11. Match Splinters

Bombax ceiba, *Pinus roxburghii*, *Ailanthus* spp., *Poplar* spp. (Safeda), etc.

12. Wood for Packing Case

Corylus colurna (Kapasi), *Mangifera indica* (Aam), *Alnus nepalensis* (Utis), *Ficus religiosa* (Pipal) *Pinus* spp., *Butea* spp. etc.

13. Walking Sticks

Zanthoxylum alatum (Timru), *Fraxinus* spp., (Ang), *Quercus himalayana*, *Bamboo* spp. (Bans, etc.), etc.

14. Rifle Parts and Gun Stock

Juglans regia, etc.

15. Tobacco Pipes

Acasia nilotica, *Dalbergia sissoo*, etc.

16. Bent Wood Articles

Celtis orientals (Kharik), *Morus alba*, *Dalbergia sissoo*, etc.

17. Boot Last and Shoes Hccls

Zizyphus jujubs (Ber), etc.

18. Brush Back

Dalbergia sissoo, *Adina cordifolia*, etc.

19. Battery Separators

Michelia champaca (Champak), *Adina cordifolia, Pinus wallichianum.*

20. Paper Pulp

Broussonetia papyrifera (Paper mulberry), *Daphne papyracea* (Satpura), *Cannabis sativa, Eucalyptus* spp., *Bamboo* spp., *Picea smithiana* (Rai), etc.

21. Tenin (From Bark, Leaves, etc.)

Acacia spp., *Quercus* spp. (Banj Moru, etc.), *Terminalia arjuna* (Arjun), *T. Chebula* (Har), *Shorea robusta, Cassia fistula* (Amaltas), *Zizyphus jujuba, Chinnamomum Tomala* (Dalchini), *Anogeissus latifolia* (Bakli), *etc.*

22. Dyes (From Wood, Leaves, Flowers and Fruits etc.)

Acacia catechu, Artocarpus integrifolia (Kathal), *indigofera* (Neel), *Lawsonia inermis* (Mehndi), *Curcuma Longa* (Haldi), *Rubia cordifolia* (Manjit), *Berberis* spp., *Terminalia tomontosa* (Sain), *Mimusops elengi* (Maulsari); *Butea frondosa, Nyctanthes arbortristis* (Harsingar), *Mallotus Phillipensis* (Rohini), etc.

23. Oil Extraction from Seeds

Brassica spp. (Rai, sersoan), *Sesamum indicum* (Til), *Prunus* spp,, *Shorea robusta, Ricinus communis* (Arandi), *Linum usitatissimum* (Alsi), *Helianthus annus* (Sooraj mookhi), etc.

The rich forest heritage of the area are not only meet requirements for fuel, fodder, timber and other forest products but also provide renewable resources of medicinal and other aromatic plants, frequently known to occur in nature and can be cultivated according to their suitability as per as soil and climatic conditions. Many important but less known medicinal plants are occurring in the unexplored area which are well known to villagers and they have been utilizing them in preparation of Ayurvedic medicines to curve many diseases. These medicinal plants in the form of herbs, roots, bark and leave etc. are extracted and taken out to other States for further processing. These plants can be cultivated or collected in systematic and planned way according to the demand of particular plants in such a way as to bring immediate economic relief to the local inhabitants. The important medicinal plants of the area are given in Table 8.4. Though many species of lower plants are of great importance but only higher plants have been described here as these are easy and convenient to identify, exploit and transport.

Although distribution of medicinal plants is mainly governed by the elevation and topographic features but a few species require special habitat with unique climatic conditions like moisture content, soil-rock type, slope, etc. The *coleus forsskohlii* which is an important perennial medicinal herb and being exploited from the study area for its commercial use and transported outside the State for processing and medicine preparation. It is observed that generally occurs between 700 and 2000 m. height and shows its luxuriant growth at steep slopes on well drained soil chiefly consisted by Quartizite rock's loose material. The poor growth is also observed along road side and grassy slopes while crad of Quartizite rocks filled with fertile soil represent healthy growth of plants. The herb has been found sum loving and contain swollen

TALBE 8.4

Important Medicinal Plants of Garhwal Himalaya

Sl. No.	Botanical Name	Local Name/ Hindi Name	Family
1.	*Aconitum nepellus*	Ateesh/Ativisa	Ranunculancease
2.	*A. fulconerii*	Vis/Ratsnam	Ranunculancease
3.	*Adhatoda vassica*	Bansa	Acanthaceae
4.	*Aegle marmelos*	Bale	Rutaceae
5.	*Atropa belladona*	Sag-angur	Solanaceae
6.	*Azadirachta indica*	Neem	Meliaceae
7.	*Bauhinia variegata*	Gooriyal/Kachnar	Caesalpiniaceae
8.	*Berberis aristata*	Kingore/Totar	Berberidacease
9.	*Betula utilis*	Bhojpatra/Birch	Betulaceae
10.	*Cannabis sativa*	Bhang/Charas	Cannabinaceae
11.	*Cinamomum Zeylanicum*	Dalchini	Lauraceae
12.	*Cassic fistula*	Shephalika/Amaltas	Caesalpiniaceae
13.	*Datura stramonium*	Dhatura	Solanacease
14.	*Dioscoria deltoidia*	Dioscoria	Dioscoreaceae
15.	*Emblica officinalis*	Amla	Euphorbiaceae
16.	*Ephedra gerardiana*	Tut-Gantha	Gentacease
17.	*Eucalyptus globulus*	Eucalyptus	Myrtacease

18. *Mentha arvensis*	Pudina/Mentha	Labiatae
19. *Myrica Nagi*	Kaphal/Katphal	Myricaceae
20. *Ocimum sanctum*	Tulsi	Labiatae
21. *Papaver somniferum*	Post/opium	Papaveraceae
22. *Plantago ovata*	Isbagale	Plantaginaceae
23. *Saussurea Iappa*	Kuth, Kushtha	Compositae
24. *Swertia chirayita*	Chirayata	Gentanacease
25. *Rauwolfia serpentina*	Sarpgandha	Apocynaceae
26. *Terminalia bellirica*	Bahera/Bevitki	Combretaceae
27. *T. Chibula*	Harad/Haritki	Combretaceae
28. *Viola odorata*	Viola/Blue violet	Violaceae
29. *Zanthoxylum alatum*	Timru/Tejbal	Rutaceae
30. *Zingiber officinalis*	Adrak/Ginger	Zingiberaceae

roots due to food storing nature. The two blocks *i.e.*, Khirsu and Kot of Pauri district are having good distribution potential of this herb.

Many more plant species have been in use for various diseases since long. These are the species of *potentilla*, *Anemone*, *Achillea*, *Calamintha*, *Crisium*, *Codonopsis*, *Delphinium*, *Geranium*, *Hyosocymus*, *Iris*, *Meconopsis*, *Origanum*, *Pedicularis*, *Podophyllum*, (Ban Khakri),*Rhodoendron* (Burans), *Spirea*, *Thymus*, *Picorhize* (Kawai), *Nardostachys* (Mansi), *vallariana* (Sumya), *Taxus*, *Acorus* (Baj), *Hydeobis* (Kachoor), *Archeodel trifolia* Salam panja), *Cedrus*, *Centrela* (Brahmi), *Gentiana* (Trapman), *Thalictrum* (Pealmola), *Plumbag* (chitrak), *Levia* (Moonj), *Vergenia* (Silfar), *Hyoaymus* (Jawan), *Trigonella* (Methi), etc.

6. Ecological Degradation of Resources due to Human Interference

The developmental strategies and their implementation have profound influence on society and environment as they provide employment, generate goods and services and are responsible for distinct rise in standard of living. Due to development of sophisticated living system the resources like forest, wildlife, mineral water and soil, etc. have gained much importance for their utilization at national and international levels, beside meeting the local needs. However, developmental process, *viz.*, logging operation, river valley projects, road construction, tourism, industrialization and urbanization, etc., also exerts pressure on the existing eco-system and has side affects, resulting into depletion of nature resources. A considerable degradation in vegetation, wild-life, mineral, soil, water and consequently overall

TABLE 8.5

Destroyed Forest Area (in Sq. Km.) for Different Developmental Activities in U.P. Hill

Period	*River Valley Project*	*Agriculture*	*Industriali-sation*	*Road Con-struction*	*Other works*	*Total*
1951-73	932.00	797.00	193.00	34.00	199.00	2,155.00
1973-74	0.41	1.78	0.62	0.78	0.13	3.72
1974-75	0.02	38.95	0.41	3.33	1.86	44.57
1975-76	3.12	0.01	—	7.86	3.75	14.74
1976-77	87.32	2.89	0.23	1.45	8.54	100.43
1977-78	—	—	—	0.69	84.03	84.72
1978-79	—	0.07	39.99	—	4.29	44.35
Total	1,022.87	840.70	48.11	234.25	301.60	2,447.53

Source : Forest Statistics, U.P. 1981, Lucknow.

environment is the net result. It is noticed that dry and deserted features are in progress resulting in multifactorial pressure on the natural vegetation as forest remain under pressure at maximum extent due to population growth and development activities.

It has been estimated that 2,447.53 Sq. Km. forest area from 1951 to 1979 has been utilized in various developmental works in U.P. hills, while the maximum area comes under the river valley project (Table 8.5).

The pine (Chir) and Oak (Banj) forests have been affected most significantly as pine is a principal source of timber wood and resin extraction while Oak fulfills the basic needs of fuel, fodder, handles agricultural implements etc. The Oak is ecologically quite significant as it has greater water retention and soil conservation capacity. In many areas the virgin forest has been converted into scrub forest due to extensive lopping by inhabitants.

A recent field check and survey of Thalisain block (Pauri district) which covers 66.70 per cent area under forest land, revealed that 13.035 Sq. Km. forest have been degraded and converted into scrub forest due to human activities like branch cuttings and tree felling, extensive lopping, browsing, etc. In addition to these, original density has been reduced upto considerable at a number of places.

It has been observed that selective felling and cutting of several important plant species have resulted into considerable change in forest composition and rarity of certain types of forest. These include *Acer caesium, Aesculus indica, Betula utilis Buxus wallichiana, Carpinus viminea*, *Juglans regia*, etc. In addition to these species many shrubs and herbs of medicinal, dye, species and other importance as *Berberis* spp., *Gentiana* spp., *Jurinea macrocephala*, *Nardostachys glandiflora*, *Rheum* sp., *Selenium* sp., *Teucrium* sp. have become undangered during the course of time. The reduced density of forest associates directly influence the undershrub and herb canopy with several hydrological and edaphic changes, lowering soil moisture and accelerating erosion. These changing habitat result in more competitive life and ultimately reduce the germination, regeneration and fruiting capabilities. Due to these abnormal conditions several terrestrial and ephiphytic forms, orchids, ferns, saprophytes and delicates lichens are finding it difficult to servive and perpetuate naturally. The development works like logging operation, road construction, etc. brought severe erosion in mountain slopes and results not only in removal of top soil layer but heavy pressure and damage to undershrub, herb and crops, etc. The environmental degradation have restricted the distribution of number of endimic plant species (Pauri 1960 and Sahni 1973) and are under great constraint. Few of the important such a endemic species are *Clarkella*, *Falcoperia*, *Hemiphragma*, *Meconopsis* *Nardostachys*, *Picrorhiza*, *Platystemma*, *Roylea* and their continued exploitation is causing threat to their perpetuation and survival.

7. Deforestation and its Consequences

The cycle of ecological ruins starts with deforestation is well established scientific truth. In addition to depletion in resources, the forest denudation resulting into deterioration of water, wildlife and other related resources and also accelerating disasterous phenomenon like land slides, soil erosion, etc. which mainly provoke flash floods.

8. Deforestation and Water Resources

The Siwalik belt which has been exploited maximum, suffered the worst revenge of deforestation. Siwalik streams have become seasonal, spring have dried up, subsoil water has gone down and agricultural productivity declined.

The field studies in the Bhabar belt of U.P. hills clearly show that the river bed have got widened about one and a half times and even the perennially flowing rivers have dried up and all this happened within the last single decade or so. Dr. Richard St. Barbe Baker, a renowned forester and man of tree in year 1978, while travelling between Hardwar and Dehradun observed the dried up streams and remarked "without going into the mountains I can now say what is happening along the sources of streams." There are many such streams in Dehra Dun and foot hill region. With unbroken regularity they swell each year with the monsoon, destroying fields and forests along their banks and convert the fertile land into waste land by spreading their gravel and cobbles. Jakhan Rao over the last 60 years, enlarged its bed from 50 m. to 500 m. or even more at places. Its water has dried up and the Khair and Shishum trees on the flood plains, have disappeared.

Drying up of fresh water springs within the hills is another major problem. In old settlements, there used to be plenty of water, have become drought prone and number of this type of villages is getting increased. The

60 km. long pipeline laid in 1966 that brought water to Pauri (headquarter of Pauri Garhwal District) is now rendered unserviceable for the source which yielded water at the rate of 360 litres per minute, can now yield only 108 litres per minute. Water will have to be pumped now for Pauri from Alaknanda river several thousand feet below.

9. Deforestation and Land Slides, Soil Erosion

The land slides is one of the major hazards in the mountaineous terrain which leads to loss of human and cattle life, their assets in the form of agricultural land, settlements, fauna and flora; it many a times creates obstruction in free movement of traffic, damages the road and blocks the turbulent streams creating huge reservoirs and leads to flash floods, Deforestation in river catchment areas is regarded as of single major factor contributing to these land slides. The frequency of slides is increased to earthquakes which are quite common due to seismic prone area. At many places, these mass movements are quite slow. This is very well observed in terraced fields on moderate slopes, where the soil cover is thin and water percolation in monsoon or due to irrigation is quite substantial. These features are originally glacial or periglacial features and the planes of separation get reactivated due to lubricating function of water and clay content of the soil. These are noticed in higher reaches (Pal and Sah, 1987). Past records the Himalaya show that from years 1808 to 1947 (139 years) the earthquakes recorded only 10 times while between 1947 to 1975 (28 years) it recorded 25 times (Arya *et al.*, 1978). Though different geological and other phenomenon like tectonism, etc. leading to environmental crises especially land slides and erosion but sharp increase in these natural calamities after 1947 (independence) indicates that the ruthless deforestation and road construction (using heavy explosives) for speedy development of the area has remained main reason. The barren hills facilitate soil erosion while cracks due to heavy explosion and geological phenomenon result into huge land slides during rainy season. Only in Chamoli district from 1971 to 1978, the 48 human lives, 949 cattles, 601 houses, costing nearly about 173 lakhs of Rupees and crop on large scale was destroyed by these mass movements. In year 1972 the Gohna lake breaking resulted a large scale destruction in Alaknanda valley and U.P. plains (Nitya Nand and C. Prasad, 1972). This flush flood engulf 10 Km. road, 6 motor bridges, 24 buses, 366 houses and 2,000 Sq. Km. peddy crop while in the year 1979, the number of villages affected by land slides was 138. Other common factors which accelerate the land slides and erosion are overgrazing, mining, seismic disturbances, the cutting by the channels, etc. The remedial measures include proper road alignment, checking of deforestation, afforestation, grazing management, scientific mining, the protection and integrated stabilisation of measures and reduction of biotic interference in slide prone areas. The soil binding plant like *Agave* spp., *Rumex* sp., *Rubus* spp. and *Berberis* spp, etc. should be planted on areas affected by soil erosion as they grow at loose soil and unconsolidated superficial deposits on slopes and bear good capability of soil binding.

The natural resources of the area are also affected by other natural calamities like cloud-brust hail, heavy snow fall, wind strom, etc. and result into loss upto considerable extent.

The resources of tropical and temperate zone have not degraded only but human interference also reached to hardly and highly accessible alpine areas. The main problem is created by nomadic Gujjara and shephards as they dwell in the forest and alpine areas throughout the year for grazing their huge herds of cattle. In addition to grazing, they indiscriminately damage the vegetation for their own requirements. Due to over and uncontrolled grazing palatable species are eliminated gradually and replaced by unwanted large, thistle like, spinescent herbs, especially *Euphorbia* spp. as well as grasses, resulting marked alternation in natural plant succession. The soil gradually become rugged by hoofs of cattle which accelerates the problems of erosion and thining vegetation, causing formation of barren patches in forest and alpine region. The shows that, till recent past the alpine meadows and pastures only favourable grazing grounds.

In Garwal Himalaya, the improved tourism, pilgrimage, increase in mountaineering expeditions, trekking, and other categories of tourism influx is also responsible upto some extent for damaging alpine meadows by their different types of activities. The damage in Himalayan environment in many places has reached a critical level and beyond the carrying capacity, therefore, in order to get out of this problem, it is the need of the time that certain more land scapes with unique types of vegetation and wild life should be conserved as "Biosphere Reserves".

Due to adverse impact on ecological succession and high spreading capability of few species. The natural vegetation is adulterated and even over laped by numerous plants of low commercial value. The *Lantana camara* (Laltain) is the best known exoitic shrub, penetrating and in many places has completely enchroached the various types of plant communities and has become problem not only for the study area but entire country. The *Parthenium* sp. which mainly grows in vicinity of human settlements causing asthma and lungs trouble through their pollen grains. The species of *Dahlia, Artemisia, Carissa, Canabis, Xanthium, Eupatorium,* etc., develop more complexion for montane vegetation. Many of the crop weeds usually introduced with the seedlings and seeds (Gaur and Nautiyal, 1980) are causing reduction in crop production. The fire caused intentionally for best fodder growth and development or accidently, lead to adverse effects on floristic composition and animal population in mountainous regions. Ultimately causes the degradation of natural environment. Due to this at several places delicate and palatable species are wiped out and fire resistant species such as *Ajuga, Artemisia, Campylotria, Carissa, Moughania, Rubus.* etc. (Paliwal and Singh, 1987) are able to exist and grass like *Imperata cylindrica, Dischanthium, annulatum Heteropogan, contortuous,* etc., become dominant in fire burned areas. One of the main adverse effects of the fires is also the total destruction of the forest litter and humus which effect the ground flora.

REFERENCES

Arya, Agarwal and Srivastava (1978). Symposium on economic and civil engineering aspects of hydrological scheme. 14-16 April, Roorkee University.

Bahuguna, S.L. (1979). The Himalaya : Towards a programme of reconstruction in "Man and Forest" ed. Gupta and Bandhu, New Delhi.

Census of India (1981). Series 22 (Supplement), U.P. Lucknow.

Champion, H.G. and Seth, S.K. (1968). A Revised Survey of the Forest Types in India, Delhi.

Duthie, J.F. (1906). Catalogue of the plants of Kumaun and of the adjacent portion of the Garhwal and Tibet, based on collection of Starchy and Winter bottom, 1846-1849. Environment of the country—DESH KA PARYAVARAN. Gandhi Peace Foundation, New Delhi, 1983.

Gaur, R.D. (1982). Dynamics of vegetation of Garhwal Himalaya. In "The vegetational wealth of Himalayas" ed. Prof. G.S. Paliwal, Delhi.

Indian Forest Policy (1957). p. 38. Ministry of Food and Agriculture, New Delhi.

Lal, A.K., Sah, M.P. Sharma, (1987). On the landuse practices in Pauri Garhwal district with emphasis on socio-economic status and degradation of land and eco-system. In Western Himalaya—Environment Problems and Development. 525-544.

Negi, P.S. (1987). Forest Resources of Surkanda Devi, Garhwal Himalaya, I.J.F., Vol. (10) 4, 283-289.

Negi, P.S. (1984). Uttrakhand Himalaya me Van Vinash, Karan aur Upai, Tehri Garhwal, U.P.

Nitya Nand and Prasad, C. (1972). "Alakhnanda tragedy—A Geomorphological appraisal." Nat. Geo. J.I. Vol. VIII, Varanasi.

Pal, D. and Sah, M.P. (1987). The role of mass movements in land degradation—a few examples from Garhwal Himalaya. In Western Himalaya Environment. Problems and Development. Ed. Pangtey and Joshi, 611-622.

Paliwal, G.S. and Singh, S.S. (1987). Some aspects and prospects of Plant conservation in Garhwal Himalaya. In Western—Himalaya—Environment, Problems and Development, Vol. II. Ed. Pangtey and Joshi, Nainital, U.P.

Puri, G S. (1960). Indian Forest ecology. Vols. I and II. New Delhi. Report of the working group on forest development and utilisation of U.P. hill area, p. 32, New Delhi, 1972-73.

Rau, M.A. (1975). High Altitude flowering plants of the Western Himalaya, Calcutta.

Sahni, K.C. (1973). Protection of endemic and relict taxa in Indian Flora. Proc. Forestry conference. F.R.I., Dehradun.

Semwal, J.K. and Gaur, R.D. (1981)—Alpine flora of Tungnath in Garhwal Himalaya J. Bombay Nat. Hist. Soc. 78, 498-512. Working plan Report of Chakrata Forest Division, 70-1971, p. 153.

9

Ecological Havoc in Himalayas : A Case Study of Kumaon Garhwal Himalayas

MISS VEENU CHADHA

The Himalayas—the King of mountains whose Scenic Splendour and awe—inspiring beauty have enthralled millions of our people since the beginning of Civilization, are to-day in peril. Man's onslaught on the resources of these mountains due to his recently acquired leverage with Science and Technology is tending to destroy much of their majestic environment.

The Kumaon-Himalaya—320 km long, is limited by the Sutlej river on the West. The best known peaks here are Nanda Devi (7,816m), Badrinath (7,069m), Kedarnath (6,940m), Trisul (7,120m), Mana (7,273m), Gangotri (6,615m) and Jaonli or Shivling (6,638m). The Southerly bifurcation here becomes the Dhauladhar range.

The Kumaon—Garhwal Himalayas comprise eight districts in Uttar Pradesh. This region has exercised a great influence on the environmental set up of northern India and the people living in the Indo-Gangetic plains. The Kumaon—Garhwal Himalayas have been facing the Wrath of angry Himalayas since the seventies and thousands of people have died due to landslides, cloud bursting, Avalances, snow storms, etc. The barren traits of land are widening day by day and the greenery is being reduced to less than four per cent of the net area.

The quantity of drinking water has gone down drastically and new resources are limited. The number of cattle far exceeds the supporting capacity of forests and pastures. The accelerated erosion of mountain slopes is due to the loss of vegetational cover, which in turn is responsible for degradation of the environment.

The bhabbar tract stands desolate and everywhere the ubiquitous lantana has taken over. The Garhwal Himalayas are dying with the sound of dynamite explosions, earth moving machinery, axes and power saws. These hills are cruelly pock-marked—scarred by the remains of what were once deoder and raunsil pines, chestnut and rhododendron and oak forests. The entire area of Garhwal, comprising five districts and covering an area of 29,968 km. of land, is now under less than 10 per cent of dense forest cover.

The ecological havoc has taken a toll not only of the quality of lives of individuals but also of the environment. Every tree that is felled now ultimately damages the life support system—the soil water and the climate—of the Indo-Gangetic plain. The Himalayan ranges are like a gigantic sponge. The forests, through roots, leaves and mosses catch rainwater and release it gently into the ground for storage in aquifers, which in turn give birth to streams and rivers. But, the enormous denudation has made the aquifers dry and the supply of the water is decreasing day by day. During the rainy spell the water simply cascades down the slopes into river valleys, carrying with it huge amount of top soil causing landslides and sheetfloods. In the dry weather, because the storage system is depleted, there are droughts.

The Dun Valley—once known for its water supply, now experiences shortage of drinking water. The water level in the Rajpur Canal has dropped by 80 per cent and in the Bijapur Canal by 40 per cent. Because of the deforestation, average temperature of Mussoorie has soared from 25.8°C (1950) to nearly 33°C (1986).

The limestone quarrying in Dehra Dun and adjoining areas has greatly polluted the environment. The mining activity is also disturbing the stability of the region and many landslides and landslips are ascribed to this cause. The mines have eroded the hillslides, destroyed leaf mould—a valuable organic fertilizer, polluted seasonal streams and caused agricultural production to plummet by 50 per cent. According to environmentalists, at one time, limestone dust particles from the quarries and factories was measured at 360 micrograms per cubic metre as against the safe level of 100. It has been observed that in Garhwal area, due to deforestation the traditional occupations like farming and dairying are disappearing. The air pollution in the Doon Valley and the Mussoorie hills is chiefly due to lime and cement factories.

The North face of the Bhagirathi river's catchment area lies in virtual ruin because of shifting cultivation and overgrazing. The Dun Valley and Garhwal are replete with examples of forest department "plantations" which have withered within two years. The survival rate for trees planted by the forest department is less than 5 per cent. No new schemes of reforestation have been introduced. Some voluntary organisations have now planted tun vimal and Silver Oak. Among the most poignant sights in the Dun Valley and Garhwal Himalayas are the plantations of kala banja shrubs—a hardy plant which is an excellent soil binder—which have been planted on barren limestone hillsides.

Garhwal was once a land of woods and meadows. The "sal" forests of the "bhabbar" rose to thickets of "oak", "chestnut", "bhimal", "semal" and "kachnar", intermingled with "bamboo", "woodfordia" and "chir pine". At higher altitudes, there was "kail", "tun" and "deodar". Higher still, appeared the "bugials", or high altitude meadows, the glaciated lakes and the fields of Himalayan flowers. The "sal" forests were mercilessly butchered during the British rule, as were the broad leafed, deciduous temperate forests. "Chir", found commercial application and this deforestation continued step by step even after independence. As a result the shape of Garhwal began to change.

The "bhabbar" tract stands desolated. Everywhere the ubiquitous "lantana" and "cactus" has taken over. The people of Pauri, Tehri, Uttarkashi and Chamoli are mostly agriculturist and the womenfolk are mostly busy in collecting fuel and fodder. As the forests retreat and the pastures are stripped, the search for fuel and fodder has become very difficult and the women sometime spend more than 12 hours in collecting fuel and fodder. The deforestation has also reduced the water supply and some of the springs have dried. With the loss of vegetation, the hill slopes have become naked and subject to more erosion and denudation. Almost every hillside is the victim of landslips and landslides. The debris load in the rivers is increasing, with much of it being attributed to landslides. The deforestation has also affected the extinction of wild life. M.N. Buch, an eminent scientist writes, "All over Garhwal one sees remnants of once noble trees, looking like the mutilated torsos of criminals whose limbs have been amputated on the orders of some cruel medieval king." The deforestation has also created unemployment, hunger and migration.

Kumaon region like the Garhwal, is facing an ecological disaster. The utter devastation of forests in the Bhagirathi catchment, including that of the tributaries such as Bal-Ganga, Bhilangana, Mandakini, etc. has created an environmental holocaust. Driving up from Kathgodam one can notice a ruthless cutting of the trees along the roadside. It is only near Nainital that some greenry is noticed. Temperate zone forests, *e.g.*, oak, seem to have disappeared. The valley slopes have become naked, barren, bare and this has a disastrous effect in lowering the water table of the region. Away from Nainital as we move to Ranikhet, Sitlakhet, Almora, Kausani and Gwaldham, we notice a wholesale tree-felling. An isolated patches of Chir, Kail and deodar are seen but for how long no body knows. Some enlightened villagers are also protecting some patches of oak. In the valley of Kosi, except some odd patches here and there, there are hardly any tree visible. The whispering pines of Kausani are absent now. An overall view of Gwaldhan is browm rather than green.

The area around Ramgarh, Sitla and Mukteshwar, once orchard heartland of Kumaon is now under the grip of ecological disaster. The disappearance of forests in this region has a direct impact on the orchards, largely because water availability has sharply reduced. The springs of the area are drying and in some, the volume has gone down. The adjoining hills in the Ranikhet, Kausani, Sitla are barren and have so little top soil that they are not amenable to immediate plantation.

Recommendations

The Kumaon-Garhwal Himalayas play a key role in the social, cultural and economic life of the people of this region. They influence the climate and contain the sources, the catchments and the watersheds of many rivers of the sub-continent, they abound in forests, plant and animal wealth, their potential for yielding energy resources, they are the home of sizeable local populations of diverse cultural heritage and richness and their attraction as centres of tourist resort and pilgrimage is growing.

Of late, the region has been undergoing considerable development. There is a paramount need to incorporate environmental and ecological considerations into the developmental activities to ensure the health of the eco-system. The present crisis facing the Kumaon Himalayas, their ecology and environment must be regarded as a challenge for the new generations and resource development. However, scattered and lacking in co-ordination, there is a growing awareness of the relationship about deforestation, erosion, energy supply and food production and of the interdependence of low-land and highland activities and their consequences. This awareness must be enlarged. The following measures should be adopted to conserve the eco-system in the Kumaon Hills.

(*a*) To save the hills from turning into a desert and alluvial plains from flash floods, a multi-dimensional project is to be prepared. For this the socio-economic constraints of the natives must be kept in view and conservation measures will have to vary from place to place and situation to situation. The first priority should be given to the process of hazard zone mapping. Once the area or belt has been identified as hazard prone, indiscriminate construction of roads, tunnels, dams, canals, etc., must be discouraged.

(*b*) A sound drainage system should be undertaken as early as possible. All streams should be diverted around the potentially hazardous area, through drains and ditches properly lined and having adequate gradient, precaution must be taken so that water does not infiltrate the ground. The bare slopes should be covered with the net of coir. The net prevents not only the erosion from the impact of rain drops but also checks disengagement of goil constituents. The netted area should be seeded with a good quality grass and applied with fertilizers. Fencing of the area must follow this which in turn should be followed by afforestation. Because the roots of the tree provide anchorage to the soil and prevents loss of water through evapotranspiration. In addition, agronomic practices should be resorted to protect agricultural fields from landslides and soil erosion.

(*c*) The plantation areas will have to be maintained for at least two decades, with gap filling, silvicultural thinning and fire protection forming part of the programme. There should be good participation of villagers in each and every programme of social forestry. There should be "Van panchayats" to protect the forests. In fact the officers of the forest department and the natives must work in perfect harmony to save the Kumaon hills from ecological disaster.

(*d*) In the degraded areas, conifers should be planted alongwith broad leafed species. Where soil and moisture conditions are extremely adverse, such desert plants as Prosopis Juliflora would both stabilise the soil and provide perhaps the finest fuel. On slopes which can not support broad leafed varieties, "Chir" pine may be planted. Moist valleys, gullies which have some soil and wind protection, slopes with a northern aspect, must be brought under bamboo, oak, chest-nut, etc. There has also to be an intermingling of shrubs which could become the main source of both fuel and fodder.

(*e*) Steps should be taken to preserve the animals, particularly the species which are facing extinction.

(*f*) The effects of various development activities on the eco-system in the region need to be monitored continually.

(*g*) Keeping in view the importance of vegetation cover, there is a dire need of coordination among the officers of the forest department, irrigation department, soil conservation and environmentalists. This co-ordination would ensure optimum social and economic returns to the natives without doing any harm to the ecology of the area.

(*h*) Hydrological studies should be undertaken to ensure the regular flow in the natural springs.

(*i*) To prevent water run off and proper use of water for irrigation, small reservoirs, specially suited to the area, should be constructed.

(*j*) Nainital, Mussoorie, Almora, Ranikhet and Kausani are tourist resorts and need adequate environmental planning. The lakes should be kept free from silt, weed and debris. The Joshimath-Badrinath pilgrim influx has opened many areas of vulnerable enviromental resources and tender beauty. The Kumaon-Garhwal region needs proper care so that this zone of "Pure nature" is not disturbed by tourist influx. Local committees can be constituted to ensure aesthetic beauty of the region.

10

Ecology and Environmental Planning in the Darjeeling Himalaya

P.K. CHAKRAVARTI

'Ecology' means the interdependence of living species with their immediate natural surroundings. The biosphere also varies with the changing scenerio. Each type of habitat has a pre-determined set of life styles that are dictated by the environmental factors like climate, soil condition, water balance, underlying rock structure and layout of the land and the like. A part of ecology, the 'eco-system', is an "interacting interdependent environmental complex" rests on a subtle balance which remains undisturbed so long as there is least interference by intruding outsiders (Smith, 1972). A slight disruption in the ecological balance somewhere sets in a chain reaction in the entire eco-system though it can adjust itself with the changing situation upto a certain level, through its own devices, and can overcome minor imbalances. If the processes of interference are repeated very often then it gets disturbed and are affected seriously. Unprecedented growth of population, almost everywhere, has invaded, encroached and disturbed the ecological setting.

'Environment' includes mainly the physical and biotic setting all around us *viz.*, land, (soil, scenic and aesthetic beauty), climate, water-bodies, minerals, netural vegetation and animal kingdom. This habitat is of immense value to mankind because the modern material civilisation is entirely based on the exploitation and utilisation of the existing resources drawn from the environment and created through human efforts. With rapid multiplication of people all over the globe, most of these natural endowments strewn around us are being constantly and ruthlessly depleted at an alarming rate caring least for the future generation. This has very aptly led to a global concern for a careful environmental planning having a conservation orientation in it.

The Darjeeling Himalaya comprising of the three hill sub-divisions (Darjeeling sadar, Kurseong and Kalimpong) is one of the least economically developed regions of West Bengal. Its backwardness can be attributed to physical handicaps like rugged terrain, harsh climate, varying altitude with steep slopes, entrenched valleys, dense forests, etc. which make living painstaking, and hazardous. The physical isolation and economic backwardness have a bearing on the social life of the hillfolk which is often ventilated through disbelief, frustration, linguistic and political autonomy. The entire landscape was peaceful and serene till 1840's. A handful of innocent and simple souls used to inhabit the sleepy villages at the backdrop of the vast panorama consisting of majestic snowcapped peaks. In order to save themselves from the gruelling heat scorching sun in the plains, the colonial rulers sent pioneers to find out a suitable site for a sanatorium in the Himalayan region. After a painstaking and arduous journey two civil servants—Capt. Lloyd and Mr. Grant—reached the village of Darjeeling in February 1829, completely exhausted at the fag end of the day. The following morning brought them good luck. At day-break, they were captivated by the exquisite scenic grandeur of the vast Himalayan mountain and invigourating climate which removed all their fatigues and hazards in a moment. They decided then and there to select this village as a site for the proposed sanatorium which was ultimately approved by the colonial rulers. Thus, through sheer accident Darjeeling was discovered and since 1835 the saga of development is going ahead. Darjeeling will be celebrating her 150th birthday next year; in the meantime she was earned international fame as a famous tourist spot and producer of high quality flavoured tea. Thanks to the zeal and initiative of the colonial rulers who discovered Darjeeling, planted tea in the surroundings (1856), laid out the Hill Cart Road (1865) and built the narrow gauge 'wonder' DHRLY (1881) in such an inhospitable terrain. These all contributed to its early start, growth and development through the three T's—tea, timber and tourism. Since then Darjeeling never looked back along with other hill stations *viz.*, Kurseong, Kalimpong, etc. but much of the old pristine glory is gone for ever.

Man has been construed as the cancer on this earth. Like cancer which affects living cells quickly and consumes the body in a similar fashion man being overzealous, acquisitive and extractive by nature always wants to dominate the environment thereby causing ecological degradation. Excessive demographic threat coupled with technological advancement have wrought incalculable damages that are apparent on a global scale. If this present trend of eco-destruction continues unabated then a major tragedy surely lies ahead.

Classification of Environmental Resources

The environmental resources of the Darjeeling Himalaya can be grouped into the following categories :

(*a*) Inexhaustible—immutable : Scenic beauty, bracing climate;

(*b*) Inexhaustible—misusable : water power;

(*c*) Depletable—maintainable—renewable : soil fertility;

(*a*) Depletable—maintainable—non-renewable : soil wash, soil creep;

(*e*) Depletable—non-maintainable—reusable : metals, non-metallic minerals;

(*f*) Depletable—non-maintainable—non-reusable : coal, Forests wildlife.

The grand scenic beauty and climate of the Darjeeling Himalaya have promoted tourism as an important industry; the local soil has aided the growth of plantation and horticultural activities; water resources have aided power generation, irrigational and potable water; forests supply raw materials to forest based industries and together with wild life provide recreational facilities; minerals, mainly coal, supply energy on a limited scale. Despite her richness in natural resource base limited planned exploitation and utilisation have taken place as yet.

Environmental Constraints

Constraints are many that are encountered for proper resource utilisation. They can be grouped as follows :

(*a*) *Environmental Constraints* :

(*i*) Rugged topography;

(*ii*) Erosion-Natural and manmade (land slip, slope failures, soil erosion, siltation, change in river courses, etc.);

(*iii*) Change in climatic regime and hydrologic balance;

(*iv*) Depletion of biospheric reserves (fuel and timber);

(*v*) Unutilised minerals;

(*vi*) High percentage of unproductive land (roads, settlements, etc.) and limited percentage of level land and

(*vii*) Defacement of aesthetic beauty through deforcstation.

(*b*) *Economic Constraints* :

(*i*) Serious ecological threat due to multiple human occupance;

(*ii*) Scacity of power/energy resources for industrial development;

(*iii*) Low transport density and network;

(*iv*) Lack of infrastructural facilities;

(*v*) High rate of unemployment—flight toward the plains.

(*c*) *Socio Cultural Constraints* :

(*i*) Physical isolation has caused backwardness and conservative outlook in remote areas;

(*ii*) Lack of skill, training and specialisation;

(*iii*) Low level of literacy in rural interiors and

(*iv*) Harsh cultural shocks mostly in urban areas, due to over rapid urbanisation and modernisation—conflict with the immigrants.

(*d*) *Special Problems* :

(*i*) Illegal immigration from Nepal creating socio-political tensions;

(*ii*) Aspirations of the indegens for political, linguistic and cultural identities;

(*iii*) Deterioration of the environmental qualities and

(*iv*) Economic frustration among the hillfolk.

In spite of so many problems the completion of the Hill Cart Road and the Darjeeling Himalayan Railway heralded a new era. Hitherto inaccessible areas were ripped open to outsiders, tea industries attracted large scale immigration labour, flourishing tourist trade further encouraged nucleation and urbanisation. With rising job opportinutes the inflow was fast and most of the job seekers were outsiders. This population invasion in huge numbers led to encroachment, eco-destruction and rural habitat transformation. The growth and development all around made excessive demand on the local resources without caring for any conservation for future generation. Environmental degradation was only natural which ensued from excessive human interferences. Lush-green sylvan canopy which once crowned the landscape has vanished yielding place to settlements, factories, tea gardens, agricultural plots, etc. The steep hill slopes are bare, devoid of any vegetal cover and have encouraged slope failures, soil wash in some vulnerable areas where old crumled rocks overlie younger rocks at the bottom. As they are exposed to rain and sun and have no tree or soil cover they are in the habit of running down the slopes because of chemical and mechanical weathering processes. These landslides cause immense hardship to the hill economy during the rainy season. Unmethodical road and building constructions allaround have also aided slope failures. Soil erosion, slope failure all carry unassorted debris and deposit them in the river beds thus choking them and leading to change in river courses, floods, etc. The vanishing green is also causing the disapperance of valuable animal species much to the annoyance of the naturalists.

Much of the green and glory of the environment in the Darjeeling Himalaya is lost. Human inroads have stripped off her beauties as well as tilted the ecological balance through large scale misuse of the natural resource base. All these misdeeds call for an urgent resource and environment planning for better habitat transformation.

Strategies for Environment Planning

The Darjeeling Himalaya today is exposed to accelerating environmental and social changes; although change is inevitable and desirable when it improves the living condition of the local inhabitants, a strategy of change must be found which is consistent with the preservation of the unique flora and fauna, spectacular natural beauty and distinctive cultures of this area (Lall, 1981).

Before the formulation of strategies for environmental planning in the Darjeeling Hill areas it is desirable to set the goals of the study.

Objectives :

(*a*) Assessment of the impact of human occupance on the environment;

(*b*) Evaluation of the ability of human groups to meet their needs;

(*c*) Analysis of the effects of changes on both man and environment through :

(*i*) Recreational change;

(*ii*) Agricultural farming and change;

(*iii*) Urban Landuse and ecological change;

(*iv*) Population and economic development;

(*v*) Other changes.

(*d*) Study the obstacles on the optimum utilisation of the resource bases (natural and human); identify constraints and strategic points for new inputs and infrastructures to stimulate the processes of development.

(*e*) Device approaches to educate the people on environmental planning and involve them in developmental programmes.

In order to achieve the desired results in regional development, employment potentials, regional income and general improvement in the quality of human life in this hilly terrain, the following processes should be involved in the formulation of an enviromental plan :

(*a*) Appraisal of resource bases (natural and human) and preparation of resource inventory;

(*b*) Identification of key problems in resource use;

(*c*) Formulation of action oriented strategies to remove the problems of the environment and exploit the development potential;

(*d*) Implementation of policy programmes through action groups;

(*e*) Evaluation of the effectiveness of the strategies and feeding back the results.

The strategy for development should incorporate a conservation oriented, integrated environment planning and resource utilisation programme. For achieving the goal the following procedures are suggested :

(*a*) Habitat transformation through the improvement in infrastructural facilities (transport, power, irrigation, employment sources, etc.);

(*b*) New landuse methods and control (ecology oriented new farming devices, crop rotation, introduction of short duration varieties of crops, horticultural development, soil enrichment programmes, etc.);

(*c*) Social and community forestry (afforestation and scientific forest management);

(*d*) Industrial development (tourist based/forest based/agro based/mineral based/animal based industries, etc.);

(*e*) Local involvement in the planning processes and

(*f*) Formulate conservation practices of resources for the posterity.

The dynamics of environmental destruction have assumed a global dimension in recent decades and aptly called for deft and immediate handling in this fragile environment. Some of the damages that have taken place already are irreversible (afforestation, an extremely lengthy process is practically impossible in bare areas devoid of any soil cover) and time consuming as well.

Continuous experiments and trial will be necessary before a new dynamic methodology is evolved. There is hardly any measure by which we can control natural erosion in the hills but some procedures can be adopted to check the constant flow of migrants, man made erosion (cultivation on steep slopes, grazing, timber extraction, fodder and fuel collection etc.) which greatly affect the climate and hydrological balance.

The entire aspect of planning the environment has a social dimension just because man is at the core of the problem. It is essentially man's influence that has upset the natural balance, and it is in the adaptation of his interaction with the environment that we must seek for ways to restore it. Unless the human behaviour in the hills is changed, all hopes of saving the rapidly deteriorating eco-system will be disappointed. Destruction of the eco-system is taking place in the hills despite the population's knowledge of the situation, and this constitutes the particular tragedy of the hills. This phenomenon can be least explained by the divergence of individual rationality from collective rationality as are expressed through the following social variants (Reiger 1981) :

(*a*) Future Ignoring Behaviour (FIB);

(*b*) Social Cost Ignoring Practices (SCIP);

(*c*) Lack of Collective Organisation (LOCO);

  (*i*) Green Apple Picking Phenomenon (GAPP);

  (*ii*) Collective Action Dilemma (CAD);

(*d*) Lack of Identification with Public Property (LIPP).

In applying these variables in the solution of social problems for effective environmental planning there is no foolproof methodology as yet. Some experiments will have to be made in a small area through pilot surveys for ascertaining the results and if successful implement in other areas.

Major Findings

(*a*) The amount of safe land for development is disappearing rapidly, while the local demand for building and road sites is growing continuously.

(*b*) As land is developed, prime agricultural land is lost and environmental quality decreases.

(*c*) Recreational demand may begin to decrease if environmental quality and infrastructural facilities deteriorate further.

REFERENCES

Chakravarti, P.K. (1983), Environmental Resources, management and Development strategies in the Darjeeling Himalaya in Singh and Others (Ed.) Volume on Environmental Management, Allahabad University.

Chakravarti, P.K. (1983), Impact of Development of the Physical Environment—A case study of the Darjeeling Himalaya—Paper presented in the National Conference on Impact of Development on Environment, Calcutta Geographical Society, Jan. 21-23, Calcutta University.

Chakravarti, P.K. (1984), Environmental Problem and Rural Habitat Transformation in the Darjeeling Himalaya in R.L. Singh and Rana P.B. Singh, Ed. Vol. IGU. Paris.

Chakravarti, P.K. (1984), Environmental Changes in the Darjeeling Himalaya : A Study in Human Ecology—UNESCO Sponsored Seminar on Man and Environment, Deptt. of Geography, Dhaka, Bangladesh, April 1984.

Dash, A.J. (1947), Bengal District Gazetteer : Darjeeling, Bengal.

Hooker, J.D. (1969), Himalayan Journal (New Ed.) Reprint, New Delhi.

Ray, B. (1967), Darjeeling—District Census Handbook, Calcutta.

Rieger, H.C. (1981), Man versus Mountain—The Destruction of the Himalayan Eco-system in Lall J.S. (Ed.) The Himalaya—Aspects of change, New Delhi.

Smith, R.L. (1972), The Ecology of Man—A Eco-system Approach. New York, p. 3.

11

The Impact of Jhum on the Eco-system of Manipur Himalaya

DR. M.L. DEWAN, C.L. BHATIA, S.N. AHUJA, B.N. SEHGAL
IN COLLABORATION WITH N.D. TYAGI, IAS, ET AL.

Jhuming or shifting cultivation is a time-honoured tradition in Manipur and the neighbouring areas of the North-East. Hill tribes have been cutting down forests to make way for agriculture and where they find it difficult to dispose of the wood after the trees have been cut down, they make to the quicker, expedient of setting fire to them and then clearing the debris for breaking the land for cultivation.

Jhuming has not only cut down the forests, hut has led to development of a more serious problem, which is soil erosion, on a large scale. Soil conservation measures undertaken in the State have failed to make any impact on the problem. A proposal to control the shifting cultivation had been formulated by the Central Ministry of Agriculture a decade back, but it had not been implemented. Contour bunding, terracing and protective afforestation have, of course, been undertaken but to a limited extent with the result that the problem remains where it was.

The committee of four experts in soil conservation, forestry, rural development and horticulture has suggested that the key to arresting degradation of upland catchment areas in Manipur like in many other parts of tropical and sub-tropical India and Asia lies in enabling farmers to establish farming practices combined with physical measures to check erosion and floods. Reforestation is only a partial solution of the problem and a broader approach is needed. There must be a farming system, which includes a strategy for fodder production, introduction of improved breeds and effective marketing systems.

The highlight of the Manipur Experiment suggested by the committee is that jhuming would be given up by the farmers only if a viable alternative is offered to them, the alternative being a system of cultivation which would give the farmer the same income that he gets from jhum cultivation. The farmer has to be involved in the development of alternative strategy.

The committee felt that the farmers would give up the jhum system only if there is an assurance from the Government that they will be able to stick on to their lands, even if Government or voluntary agencies decide to plant trees on them. Since the Government agencies have not been able to do any thing in the direction of soil conservation, it is vital that voluntary agencies be involved in the endeavour.

The essence of the committee's recommendations in regard to getting the cooperation of the people is that we must go to the people, live with them, learn from them, serve them, plan with them, start with what they know and build on what they have. The Committee was offered three sites in Ukhrul district, Chandel district and Khamshin village (Chandel district). It was found that the people had more faith in financial institutions and voluntary agencies than in the official paraphernalia.

Kamjong Agro-forestry Project

One of the three sites selected by the Committee was the Kamjong area in Ukhrul district. The project envisages that an area of 40 hectares would be brought under fruit trees including citrus fruits. The Committee offered a scheme whereby one hectare of land would give an income of Rs. 18,000 per year. A 20-hectare plot of land is proposed to be planted with bamboo, poplar, eucalyptus, oak, perkia rexburgaiperkia jawanica, willow and bakain trees. NABARD has already approved a scheme for poplar plantation in the northern plains which is being executed through M/s. Wimco Seedlings Ltd. under which 1,000 trees are to be planted per hectare which will give Rs. 500/- per tree after eight years. The farmer is paid a part of the cost of plantation. Seedlings and technical know-how to the farmers is provided by the company. It is felt that a scheme for the whole State can be formulated which will provide an alternative means of livelihood to those who are engaged in jhum cultivation.

The Committee has formulated detailed schemes for the plantation of trees of various varities which will not only provide an alternative means of livelihood to the farmers but would also provide the necessary forest cover for the conservation of the soil which is being depleted at a fast rate because of the jhum system of cultivation.

Schemes for the plantation of coffee, cardamom, turmeric, rubber, pepper, ginger and cinamon would also be prepared in consultation with the banks which would provide the necessary finance to the farmers to be involved in the schemes.

Chandel Christian Village Project

The Committee formulated another scheme for the Chandel Christian Village whose chief agreed to give five hectares of land for the project. The project is an agro-forestry project under which bamboos, poplars, eucalyptus trees and plantations crops like cardamom and coffee would be planted. The project also envisages the plantation of fruit trees like pineapple, pears and plums, walnut and almonds, papayas and bananas and exotic trees like mangestian and rambutan.

Khangshim Village Project

The third project recommended by the Committee is the one to be undertaken in the Khangshim village in Chandel district. Here land measuring 5 hectares is proposed to be brought under tree plantation for which the banks would be giving assistance at the rate of Rs. 2,000 per hectare. The plantation would also serve as a training centre for the farmers of the area around the project.

Imphal River Training Programme

If the Committee's recommendation is accepted, a ten kilometre length of the Imphal river bank would be taken up for plantation of trees on both the banks. It would not only beautify the area but would also act as a conservation measure. People's nurseries would be developed so that there is no dearth of seedlings.

The river training programme would also take care of the pollution of the waters which flow through the town. An offshoot of the programme would be the establishment of school nurseries on the Gujarat pattern. A look at the Gujarat School Nurseries Programme undertaken with the help of the World Bank in 1980 would convince anyone of the viability of the venture. A similar approach could be adopted in Manipur. In Gujarat the school nurseries were able to garner an income of about Rs. 38 lakhs in four years. The amount was utilised by the schools for educational tours, games kits and other purposes. There is reason to believe that if a similar programme is undertaken in Manipur the State would benefit to a considerable extent.

Orchid Production

The Committee has also recommended that orchid production be undertaken in the State on a commercial scale. It would lead to generation of income and the development of export of cut flowers and flower seedlings. It has been recommended that school boys and girls be trained in orchid growing.

The Committee has recommended the establishment of an infrastructure through the setting up of a co-operative society to manage the various projects underlined above. The primary objective of the society should be to undertake the various projects for income generation for the farmers now engaged in jhum cultivation. Once an alternative is provided to them they would gladly give up the shifting type of agriculture which is leading to soil erosion on a large scale.

12

The Impact of Urbanisation on the Eco-system of Himalayas

Y.P. Talwar

The agroclimatic patterns in our country vary from dense tropical forests with a precipitation around 2,500 mm to cold deserts at sub-zero temperatures with meagre vegetation. The entire range of rainfall from over 2,500 mm to less than 500 mm is well represented in the sub-continent. While some of the problems afflicting those lands are inherent, some are the result of increasing human and cattle needs, centuries of conservative agricultural practices and geomorphic evolution. On the one hand the rural inhabitants have to face problems of soil erosion, flash floods, droughts, avalanches, typhoons, storms, etc., and on the other rising population. Increasing pressure of human and animal needs, rapid denudation, biotic interference in natural regeneration have contributed to increasing influx of humanity to urban areas. The problem of urbanization has created ecological imbalance.

Ecological problems embrace diverse aspects, ranging from the economic, social and psychological problems of human settlements to the management and use of natural habitats. The environmental degradation of developing cities in the Indo-Gangetic Plain and along the foot hills of Himalayas are not the side effects of excessive economic depressions and industrialisation, they reflect the inadequate planning and development.

Industrialization brought about mass production techniques, mechanization and diversification of work and these, along with improved agricultural techniques responsible for creating food surpluses, were responsible for the rapid population increase in the sub-continent. It is said that the larger the size of population, the greater are the possibilities of division of labour. And thus the lure and temptation of a variety of avenues of employment in the urban areas—which are recognized as centres, not only of industry, trade and commerce, but also of education, culture and the arts—has unleashed the demographic flood into the cities. It has been estimated that the population of slum and squatter settlements in urban settlement areas as increasing at twice the rates of growth of population of the cities. This is four times faster than the world population growth.

In terms of growth of urbanized human settlements, India is by no means a highly urbanised country. According to 1981 census about 24 per cent of the total population lives in urban areas. In absolute numbers, the total urban population living in urban settlements comes to 156 million which is large enough by any standards.

This is creating a rural urban divide resulting in inter-regional and inter-personal differences. This is also leading to deteriorate the standard or quality of life, environmental imbalance, environmental degradation, depletion of resources, breaking down of services and environmental pollution.

The basic need for people in any human settlement is housing, basic amenities such as fuel, fodder, water etc., when the population in a particular area goes on increasing in an unplanned, unregulated manner and the water supplies available are severely limited, the obvious result is the decline in the quality and quantity of the water supply to the consumer. This results in a multifold increase in the incidence of water borne and water based diseases.

Again, our wasteful life-styles generate large quantities of garbage. The garbage dumped on roads, streets, etc., is disposed of after few days by the municipal workers and this has resulted into the emergence of many diseases. Even in small town settlements the discharge of garbage has become a serious health hazard.

It is heartening to note that in Delhi, the landfill with garbage has been utilised for electricity generation. Thus, garbage waste can be usefully utilised for becoming an environmental asset in the urban settlements. However, in rural areas the biogas plant can be installed for generating cooking gas fuel.

The rising data of migrants from villages to the cities has led to the emergence of slum and squatter settlements. The lack of latrine facilities and its maintenance has resulted into a stinking environment indicating the unhygienic way of life. Thus in slum and squatter settlements there is a serious deterioration in the quality of life.

The density of residential areas in the congested cities of the outer Himalayas has been rising and causing health hazards. A large number of people in the congested metropolitan and large cities are living in the room tenement. This has, however, created an overcrowding situation in urban settlements with the denial of fresh air, light and privacy.

The health services has greatly deteriorated due to the population explosion. The programme of family planning has not been very successful in the rural areas of Himalayas. As a result, the rural areas are experiencing the population chaos of starvation, hunger, poverty and unemployment. The population pressure has a great constraint on land resources and causing traffic and transportation problems which in turn result into environmental pollution due to emmission of vehicle smoke. Due to the rapid and excessive urbanisation there has arisen the large scale industrial development causing a serious deterioration in the quality of life.

As a result of urbanisation, the pressure on land has led to the swallowing up of any and every kind of open space in the name of building apartments and office complexes. The deforestation in the newly developing colonies in the Himalaya has greatly affected the eco-system. And if the present urban eco-crisis goes unheeded and unchecked, humanity has to face problems of gigantic proportions.

The environmental challenges need to be regulated through appropriate tools of legislation. Ineffective environmental legislation has aggravated the rising air and water pollutions. The urbanisation has to be checked and discouraged. There should be a proper policy matter regarding the settlements, proper land use, growth with social justice, employment opportunities and curbing of migration to high density metropolitan areas.

The national and central financial institutions and banking structures should be involved more actively in the process of lending funds to the farmers, agriculturists, small establishments, craftsmen, handloom workers, carpenters, etc., in the rural areas to minimise the chances of migration of the village folk to the newly developed towns and industrial complexes. Employment opportunities have be created in the rural areas, new industrial units should be introduced in the sub-urban areas and projects of national importance should also be started in the countryside to curb the migration of people to the rural areas. People's participation should have a meaningful orientation concerning the urbanisation.

13

Tourism And its Impact on the Eco-System of the Himalayas

JAWAHAR LAL RAINA

Introduction

The Himalayas, the highest mountain system in the world, contain most of the world's "eight-thousands" Peaks. They are also the world's youngest and longest East-West mountain system, extending almost uninterruptedly for a distance of 2,500 km. and covering about 5,00,000 sq. km. The Himalayas have been one of the dominant features of India. They are the abode of the snow and consequently they have long been known as the Himavan, Himadri or the Himalaya. They are the highest folded mountains of the earth, rising to over 8,000 metres from the sea—level, which run in an East-West direction along the entire Northern boundary of India. They are 150 to 400 km. broad. Their areal stretch is between the Indus river and the Brahamputra, encompassing the entire Jammu and Kashmir; the Himachal Pradesh; the Dehra Dun District and Kumaon division of U.P. (the two Garhwals, Uttar Kashi, Chamoli and Pithoragarh); Nepal, Sikkim and Bhutan; the Darjeeling district of West Bengal; the States of Assam, Manipur, Tripura, Nagaland, Meghalaya, Mizoram and Arunachal Pradesh. The so-called invincible Himalayan wall is no longer an impenetrable rampart in view of the modern developments of war strategy including nuclear weapons, its geographic significance, however, remain intact.

The longitudinal sections of the Himalayas consists of the three zones of parallel ranges, *viz.*, (*a*) The Greater Himalayas, (*b*) The Lesser Himalayas, (*c*) The Outer Himalayas.

The Greater Himalayas

The Greater Himalayas were called "Bahirgiri" by the ancients. They comprise the northern-most ranges rising to an average hight of about 6,000 metres with breadth ranging from 120 to 190 km. This great mountain are terminates at both its western and eartern ends where there are sharp syntaxial bends. They are eternally snow-bound, manificient and majestic in their virgin whiteness. Some of the highest Peaks situated in this range are Mount Everest (8,848 m.), Kanchanjanga (8,598 m.), Dhaulagiri (8,172 m.), Nanda Devi (7,817 m.), Nanga Parbat (8,126 m.), etc. Some of these Peaks are very sacred not only to the Indians but also to the Nepalese and the Tibetans.

The Lesser Himalaya

The Lesser Himalaya known as "Antagiri" among the ancients, are an intricate system having an average elevation of about 3,500 to 5,000 metres with an average width of 60 to 80 km. Important ranges included are the Dhauladhar, the Pir Panjal, Nag Tiba, Mahabharat Range and the Mussoorie Range. These are also snow-capped and majestic but more friedly to human contacts and less inaccessible to Pilgrims, explorers and tourists. Famous hill resorts like Shimla, Rani-Khet, Mussoorie, Nainital, Darjeeling—all lie in them.

The Outer Himalaya

The Outer Himalaya or the Siwaliks, called "Upagiri" by theancients, consists of the foot-hills which run almost from Patwar Plateau to Brahmputra valley. They are 1,000 to 1,500 metres high with a width ranging from 15 to 50 km. It is a chain of low—lying hills, entirely made of fluvial deposits like sand, clay and rounded stones, slates, etc. The region is mostly ill-drained (like that of Terai), but cool and finely wooded, extending a friendly welcome to human effort and habitation.

Significance

The geographical feature which dominates India most is the Himalaya. There are no mountain ranges anywhere in the world which have contributed so much to shape the life of the country as the Himalayas have in respect of India. It is not only the political life of the people of India, but also the religion, the mythology, art and literature of the Hindus that hear the imprint of the great mountain barrier. In a very special measure the Himalaya is India's National mountain as the Ganga is its National river.

The Himalayan landscape with its majestic snow-clad peaks, vast ice-fields above the heads of large valley glaciers, roaring water falls over high precipices, deep river gorges and broad valleys is well represented in Jammu and Kashmir and Himachal Pradesh, Northern Regions of Uttar Pradesh, West Bengal and Assam. Jammu and Kashmir's fascination is centred on the Kashmir valley, through which the Jhelum meanders amidst lakes, orchards, rice-fields and densely populated settlements. Many picturesque spots occur in Bhaderwah and Kishtwar in Jammu and Kashmir, and in the valley of Baspa, Kulu and Kangra in Himachal Pradesh and in the Nanda Devi Group of mountains in Uttar Pradesh. A magnificient view of the snow-covered Kanchanjunga is obtained from Darjeeling in West Bengal. The Himalaya exercise as dominating influence on the meteorologic conditions of India as over its Physical geography, vitally affecting its air and water circulation systems and through these, the distribution of life. The high snowy ranges have a moderating influence on the temperature and humidity of northern India. When the neighbouring lands are suffering from scorching heat in summer, the lower and upper ranges of the Himalayas, because of their height, enjoy a very cool and pleasant climate. Owing to intense heat in the plains, India has developed a number of hill stations, especially on the Siwaliks, which lie at about 2,000-2,500 metres. The important ones are : Almora, Bhowali, Dalhousie, Dharamshala, Dehra Dun, Gulmarg, Pahalgam, Kalimpong, Shimla, Mussoorie, Solan, Chail, Nainital, Ranikhet, Kasauli, etc.

Impact of Tourism on Environment

India is a land of lofty mountains and mighty rivers. No other country in the world is richer in scenic grandeur and the panorama of contrasting land-scapes at different stages of evolution. Tourism is one of the important industries of India. Himalayas offer the tourist a variety of scenic beauty. To anglers Himalaya is a Paradise, to hikers it is a challenge. Considering the vastness of the Himalayas, its lustgreen valleys, emerald meadows, towering peaks, vast ice-fields above the heads of the large valley glaciers and rocky slopes, the number of tourists visiting the hill resorts of Himalayas, is not large enough to make a major impact on the environment. But there is a sufficient concentration of the visitors in some of the most popular resorts of the Himalayas *e.g.*, Pahalgam and Gulmarg in Jammu and Kashmir, Shimla, Kangra, Dalhousie, etc., in Himachal Pradesh. These hill resorts look like slums during the peak summer season. To accommodate the influx of the tourists, hundreds of new buildings are constructed every year/season and this destroys the natural landscape. A variety of sports for tourists of every age and temperament are offered by the State Govt. Fishing, Treking, Skatting, Boat Race, Pony Riding, Golf are some of the famous sports liked by the tourists. The fast growing trends in technological prospects leading to dramatic changes in socio-economic struture of the societies accelerates the tourism. The mass tourism demands environmental resources in a big way and with the problem of mass tourism arises the issue of protecting and conserving the environmental resources which the tourists in their wanton and holiday making moods erode often unintentionally. It is necessary that while planning developmental works due consideration should be paid to the fragile beauty and natural surrounding of that area. No space action should affect the ecological balance of that area. Any tourist activity that is permitted has to be subservient to and in consonance with the principle of conservation of nautre and of the genetic and natural resources within it. But there have been blatant violations of this basic code. For example, the establishment of Golf course in the city forest park by the government of Jammu and Kashmir. All the noise made by the naturalists failed to stop the project. Unplanned urban development along the length of our mountains is resulting is irrepairable loss. The mountains and forests, the most beautiful gifts of nature to man are being destroyed indiscriminately. The tourists intentionally or unintentionally create undesirable impact and introduce a disruptive influence on local traditions and social norms. The problem of litter, noise, erosion, fauna and flora needs prior attention. Tourists season is usually in hot summer months when hot winds strike the plains of the country and people rush to hill stations of the Himalayas to get relief from the scortching heat. Tourists also come in large numbers in the winter seasons. Tourists come to enjoy the snow fall, skatting and other winter sports. The increased number of tourists require more and more number of vehicles. The vehicles pollute the atmosphere. The hordes of tourists in snazzy tourist coaches, lumbering ambassadors and little Marutis pollute the atmosphere in the health resorts. The vehicles emit carbon dioxide, carbon monoxiode, sulphur dioxide, hydro-carbons, etc. These gases causes tuberculosis, lung cancer, asthama, etc. The garbage pollution in the hill resorts spread many diseases.

The great rush of tourists destroys the eco-system. The major source of water pollution is the human excreta. The refuse gets mingled with local streams, lakes, ponds, etc. The water gets polluted and causes cholera, typhoid, dysentry and many other diseases. It needs proper attention and proper sewer management. The State of Jammu and Kashmir is economically of great importance due to its great potential for recreation and tourism. Nearly one third of the Kashmir valley is occupied by water surfaces. In terms of tourist attraction the principal water body of the valley is the Dal Lake. All the House Boats in the lake accommodate thousands of tourists every year, the refuge directly intermingles with the water of the Dal, thereby polluting the water of the lake to a great extent. Proper steps should be taken to conserve the ecological system of the water-bodies of the Himalayas. All the lakes of the Himalayas should be saved from the pollution hazards. Often the tourists in the hill resorts lack the discipline to observe the rules and regulations. It is virtually impossible to keep track of the tourists. Local people are often frightened and panicked over the strange and sudden unaccustomed sounds and movements of tourists. Tourists trample across restricted areas in the forests. disturbing the wild life and endangering their own lives also.

The remedy to all these problems does not lie in putting a ceiling on the number of visitors to a p rticular health resort. The remedy does not lie in restricting the movements of trekkers or holiday makers. The remedy lies in the better management and proper planning of the tourist industry of each State. The tourism departments of the various States should try and divert the traffic to lesser known and less crowed areas. A limit should be put on the accommodation problem so that over crowding and erosion of the environment is not permitted. Due attention should be given to the supply of fresh water to the tourists, disposal of refuge, garbage, sanitary conditions, etc. The tourism industry should go hand in hand in protecting and conserving the environment.

There is no doubt that great damage has been due to the ecology of the Himalayas, but most of it has been done by other forces also and not by tourists alone. Dehradun—Mussoorie belt, one of the premier tourists hill resorts of India, is faced with a number of environmental problems stemning from exploitation of mineral and forest resources. The negative effects of exploitation are responsible for the loss of soil and forest reserves affecting the mountain eco-system, pollution of water and air, siltation of rivers and destabilization of hill slopes leading to land slides. The Darjeeling tea is famous the world over for its rich flavour. Its rich qualities are controlled by over 400 organic complexes, Conversely it is highly sensitive to environmental changes. The proposed super thermal plants of 2100 m.w capacity at Farakka and Kahalgaon will generate environmental pollution by way of damage caused to the tea plantation through physiological and morphological damages, damages to soil pollution of water, etc. Environmental problems are growing worse and worse with the passage of time. Soils are getting rapidly degraded, forests are disappearing leading to desertification and to floods and excessive accumulation of silts in the reservoirs. If these trends continue, the world in the year 2000 will be more crowded, more pollutted, less stable ecologically and more vulnerable to disruption; than the world in which we live in.

There is a vast scope of development in Tourism. While many places had since been developed as tourists sports, many more remain un-developed. Due to lack of well developed means of transportation, scarcity of fuel and fodder and availability of the suitable accommodation in the hill resorts. Master plans should be prepared and new areas developed to meet the ever great demand of tourists resorts. The Government of Jammu and Kashmir is very much interested to boost up this industry. The tourism department is busy in developing the tourists spots in Jammu and Kashmir State. In Ladakh there are many such places which if developed can attract many visitors. The Government of India should take urgent steps to conserve the ecological system of Himalayas. A great deal is at stake and half-hearted measures are not going to give any results.

REFERENCES

Chadha, S.K. 1987, *Himachal Himalaya Ecology and Environment.* Publisher, M/s Today and Tomarrow's Printers, New Delhi, pp. 121-128.

Chatterjee, S. P. 1968, *India : A Physical Geography*, Publication Division, Ministry of information and Broad Casting, Government of India, New Delhi, p. 6.

Panikkar, K. M. 1955, *Geographical Factors in Indian History*, p. 33

Radha Krishna, B.P. and Ramcharan, K. K. 1968, *India's Environment Problems and Perspectives*, Geological Society of India, Bangalore, pp. 235-239.

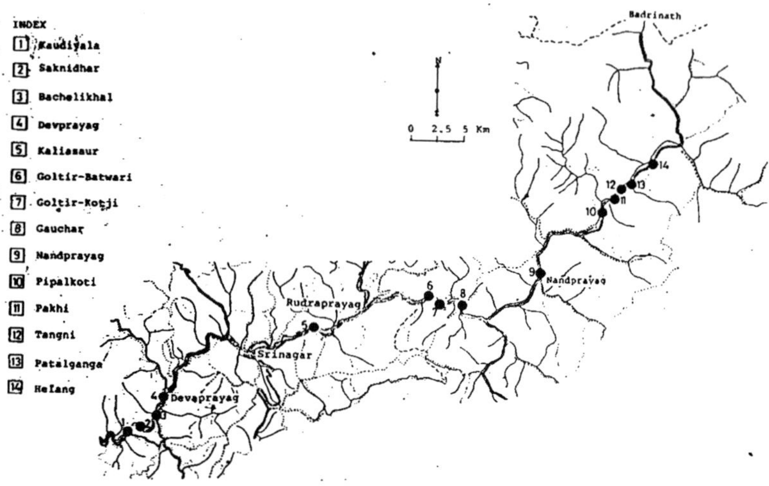

Landslide areas surveyed on Hardwar-Badrinath Road

(Courtesy Shri MAN MOHAN)

The crushing burden imposed by deforestation

(Courtesy Shri M.N. Buch)

Tehri project site: seismic zone

Boom or Bust? Excessive tourism is destroying Nainital's ecology

:w of the Glaciers in the Zanskar Range. The
.iment of some of the vital rivers of India

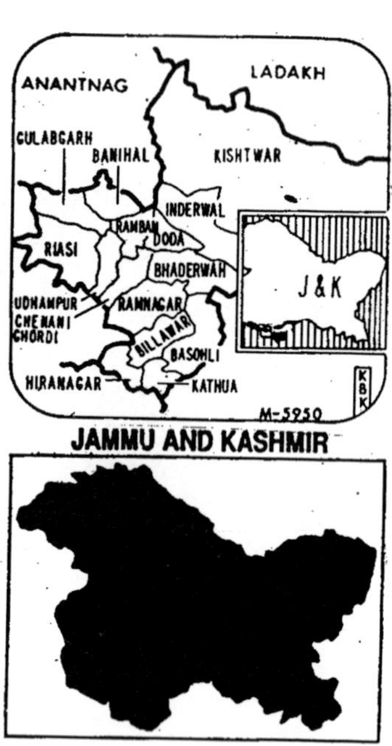

The political boundaries of Kashmir

A view of the Snow Covered Ladakh range with a glacial lake at the base

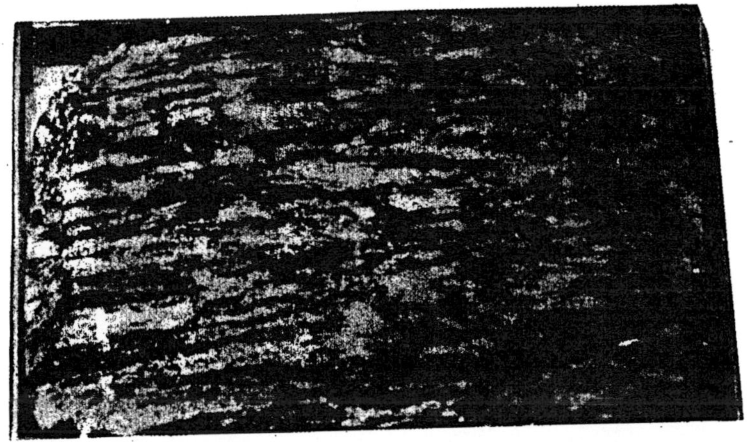

Gneissic rocks of Ladhak containing plagioclase, quartz & biotite. The rocks belong to the Salkhala Formation of pre-Cambrian age

The Indus Gorge with river flowing at the bottom between steeply rising banks, Himalayas

A view of Sonamarg (Kashmir) – An interesting area for visitors

A Panoramic view of Pahalgam (Kashmir)

A view of the Nainital Lake – A hill resort and a tourist attraction

A huge bin in a children's park at New Rajinder Nagar in New Delhi

Express photograph by R.L. Chopra